UNECE

Guidance Note for the Implementation of Source-to-Sea Management in Transboundary Basins

UNITED NATIONS

Geneva, 2025

Note

The findings, interpretations and conclusions expressed herein are those of the author(s) and do not necessarily reflect the views of the United Nations or its officials or Member States.

Links contained in the present publication are provided for the convenience of the reader and are correct at the time of issue. The United Nations takes no responsibility for the continued accuracy of that information or for the content of any external website.

The designations employed and the presentation of the material do not imply the expression of any opinion whatsoever on the part of the Secretariat of the United Nations concerning the legal status of any country, territory, city or area, or of its authorities, or concerning the delimitation of its frontiers or boundaries.

This publication is issued in English, French, Russian and Spanish.

Symbols of United Nations documents are composed of capital letters combined with figures. Mention of such a symbol indicates a reference to a United Nations document.

Photos for this publication were provided by: Winnifred Akiso, Ruth Mathews, Melisa Wandia, Jurgo Van Wyk, Sam Weru.

Cover photo: © Krasimir / Adobe Stock (for all other photos featured).

ECE/MP.WAT/78

UNITED NATIONS PUBLICATION
Sales No.: E.25.II.E.16
ISBN: 978-92-1-157910-9
eISBN: 978-92-1-154488-6

Foreword

Water knows no borders. From freshwater sources to the coast and out to the sea, water connects us all. Yet too often, our approaches to its governance remain divided. Freshwater and marine communities work in silos, and upstream and downstream actors are disconnected. Land, freshwater, coasts and the ocean are treated as separate parts instead of one system.

Rivers, lakes, wetlands and ocean ecosystems are inextricably linked, as is their health. When we trace the sources of marine pollution, we find that around 80 per cent originates from land-based sources – from our farmlands, our cities and our industries – and flows downstream into the ocean through rivers. Infrastructure such as dams affects water flows, sediment transport and fish migration. Degraded wetlands and coastal areas weaken our resilience to climate change and reduce vital carbon storage.

These challenges also impact livelihoods, public health, food and water security, and economic development. For instance, polluted waters harm fisheries and agriculture, affecting both local economies and nutrition.

We must bridge the silos in governance, policy development, administrative approaches and practical solutions, and embrace integrated source-to-sea management, as highlighted in the United Nations Ocean and United Nations Water conferences. Source-to-sea management supports not only ecological integrity, but also stronger, more resilient and prosperous societies.

The *Guidance Note for the Implementation of Source-to-Sea Management in Transboundary Basins* is an important step forward. It builds on the fact that with over 60 per cent of freshwater flows crossing national borders, transboundary water cooperation is not optional, but essential. Plastic and nutrient pollution alone, much of which travels through transboundary rivers, illustrates the scale of the challenge and the necessity of collective action.

The Water Convention, hosted by the United Nations Economic Commission for Europe (UNECE), provides a robust global legal and intergovernmental platform for cooperation across national borders and sectors, including for the effective implementation of source-to-sea management. It promotes inclusive dialogue, collaborative decision-making and the shared stewardship of our precious water resources. Through this Guidance Note – developed under the leadership of Estonia, Panama and Slovenia – we aim to share practical knowledge and implement the source-to-sea approach in shared waters around the world.

By supporting, through structured advice and concrete examples, cooperation between freshwater and marine communities, the Guidance Note strengthens the resilience of transboundary river basins and their recipient waters by promoting more coherent governance and interventions, supporting sustainable development, and ultimately contributing to stability and peace in international river basins and beyond. It is intended to inspire and inform decision-makers, basin authorities, administrations, experts and practitioners working to strengthen cooperation and sustainability from source to sea in shared basins.

We call on governments, river basin organizations, bilateral and multilateral commissions, regional seas commissions and action plans, civil society and the private sector to embrace this approach and to collaborate across sectors and borders. This means turning shared principles into concrete action – strengthening dialogue, aligning policies, management approaches and investments, and jointly addressing pressures from source to sea.

Under the Water Convention, we will continue to support the application of source-to-sea management in transboundary basins – through capacity-building processes, exchange of practices and experiences, development of tools and project implementation, and other related efforts.

Water knows no borders. Neither should our solutions.

Kuldar Leis,
Minister of Infrastructure,
Estonia

Juan Carlos Navarro,
Minister of Environment,
Panama

Jože Novak,
Minister of Natural
Resources and Spatial
Planning, Slovenia

Tatiana Molcean,
Executive Secretary,
United Nations Economic
Commission for Europe

Preface

This *Guidance Note for the Implementation of Source-to-Sea Management in Transboundary Basins* has been developed under the programme of work of the Water Convention for 2025–2027 (ECE/MP.WAT/75/Add.1), as part of programme area 3.5: *Source-to-Sea – Linking Governance and Management of Transboundary Freshwaters and the Coastal and Marine Environment.* It reflects growing recognition of the need for holistic management of land, freshwater, coastal and marine environments, especially in basins shared by two or more countries.

The preparation of this Guidance Note is a direct outcome of the Global Workshop on Source-to-Sea Management, held on 14–16 December 2022 in Geneva within the framework of the Water Convention, under the leadership of Estonia and Slovenia, and in cooperation with the UN Environment Mediterranean Action Plan (UNEP MAP), the Intergovernmental Oceanographic Commission of the United Nations Educational, Scientific and Cultural Organization (IOC-UNESCO), the Stockholm International Water Institute (SIWI), the Action Platform for Source-to-Sea Management (S2S Platform), the Global Water Partnership (GWP) and the Global Environment Facility (GEF) and its International Waters Learning Exchange and Resource Network (GEF IW:Learn). The workshop brought together a wide range of stakeholders to explore how source-to-sea management can address pressing environmental and governance challenges across connected ecosystems.

Subsequently, the Working Group on Integrated Water Resources Management at its 18th meeting (Geneva, 19–21 June 2023), entrusted the secretariat, together with the co-lead Parties Estonia and Slovenia, with the development of a Guidance Note on source-to-sea management and encouraged partners to cooperate on this effort.

The Guidance Note aims to support a better understanding of source-to-sea management and offers practical recommendations for its application. It outlines how this approach can enhance cooperation between upstream and downstream actors, promote integrated management of natural resources and strengthen ecosystem health in transboundary basins. By emphasizing the linkages between land, rivers, lakes, aquifers, coasts and the ocean, the Guidance Note contributes to more holistic, coordinated and sustainable governance, while at the same time promoting sustainable development and economic growth, and facilitating peace.

The intended audience includes government officials, river basin authorities, Regional Seas Commissions and Action Plans, and technical practitioners working at local, national and international levels. It is especially relevant for those involved in the management of shared waters and connected ecosystems, including deltas, estuaries and marine environments.

This document was prepared within the framework of the Water Convention, under the leadership of Estonia, Panama and Slovenia. The process was supported by a lead author, the Water Convention Secretariat, and a diverse drafting group of experts from governments, academic institutions, partner organizations, joint bodies and regional sea conventions and action plans. The group presented an initial outline of the Guidance Note at the 19th meeting of the Working Group on Integrated Water Resources Management (Geneva, 6–8 May 2024).

At the 10th Session of the Meeting of the Parties to the Water Convention (Ljubljana, 23–25 October 2024), the annotated outline was presented. The Meeting of the Parties entrusted the Secretariat, in consultation with the co-lead Parties and drafting group, with finalizing the Guidance Note based on the annotated outline and received feedback, with publication envisioned for 2025.

The Guidance Note was introduced during important global events, including a side event at the 3rd United Nations Ocean Conference (Nice, France, 9–13 June 2025) and at the 20th meeting of the Working Group on Integrated Water Resources Management (Geneva, 13–15 October 2025).

Acknowledgements

Bringing this publication to life required the expertise, dedication and collaboration of a broad range of partners – from national governments and international organizations to Regional Seas Commissions and Action Plans, river basin organizations and individual experts.

This *Guidance Note for the Implementation of Source-to-Sea Management in Transboundary Basins* was developed under the leadership of Estonia, Panama and Slovenia. Their guidance and contributions were instrumental in the development of this publication.

Special recognition goes to the lead author, **Ruth Mathews** (Currents Converge), for her expertise and dedication in shaping this publication. We also acknowledge the valuable contributions of **José Murillo** (Stockholm International Water Institute – SIWI), **Erol Cavus** (United Nations Environment Programme – Mediterranean Action Plan – UNEP-MAP), and **Susanne Schmeier** (IHE Delft and Scientific and Technical Advisory Panel (STAP) of the Global Environment Facility (GEF)) who authored specific sections of the text.

The Secretariat particularly acknowledges the following members of the drafting group, whose collective efforts made the development of this Guidance Note possible:

- **Aleš Bizjak** (Ministry of Natural Resources and Spatial Planning, Slovenia, Chair of the Bureau of the Water Convention in 2025–2027 and co-Chair of the drafting group)
- **Harry Liiv** (Ministry of Climate, Estonia, Vice-Chair of the Bureau of the Water Convention in 2025–2027 and co-Chair of the drafting group)
- **Karima Lince** (Ministry of Environment, Panama, co-Chair of the drafting group)
- **Esteban Alonso Arias Carvajal** and **Giovanna Valverde Stark** (Ministry of Foreign Affairs and Worship, Costa Rica)
- **Turo Hjerppe** (Ministry of Environment, Finland)
- **Landing Bojang** (Ministry of Fisheries, Water Resources and National Assembly Matters, The Gambia)
- **Shishani Munyika** (Ministry of Agriculture, Fisheries, Water and Land Reform, Namibia)
- **Frank Zhang** (Swedish Agency for Marine and Water Management – SwAM)
- **Iryna Makarenko** (Black Sea Commission – BSC)
- **Ruth Mathews** (Currents Converge)
- **Lucilla Minelli**, **Lorenzo Galbiati** and **Marco Sacchetta** (Food and Agriculture Organization of the United Nations – FAO)
- **Astrid Hillers** (Global Environment Facility – GEF)
- **Dimitris Faloutsos** (Global Water Partnership – GWP)
- **Lotta Ruokanen** (Helsinki Commission – HELCOM)
- **Susanne Schmeier** (IHE Delft and Scientific and Technical Advisory Panel (STAP) of the Global Environment Facility (GEF))
- **Birgit Vogel** (International Commission for the Protection of the Danube River – ICPDR)
- **James Dalton**, **Maria Lindelien** and **Davison Saruchera** (International Union for Conservation of Nature – IUCN)
- **Claudette Spiteri** and **Konstantina Toli** (International Waters Learning Exchange and Resource Network – IW:LEARN)
- **Giuseppe Nerilli** (Italian Agency for Development Cooperation – AICS)
- **José Murillo** (Stockholm International Water Institute – SIWI)
- **Hartwig Kremer**, **Alex Pires** and **Stephanie Vanderpoel** (United Nations Environment Programme – UNEP)
- **Erol Cavus** (United Nations Environment Programme Mediterranean Action Plan – UNEP-MAP)

The Secretariat also wishes to thank the contributors who provided case studies to the Guidance Note:

- **Shishani Munyika** (Ministry of Agriculture, Fisheries, Water and Land Reform, Namibia)
- **Karima Lince** (Ministry of Environment, Panama)
- **Maria Apostolova** (Amazon Cooperation Treaty Organization – ACTO)
- **Raúl Ernesto Artiga Colato** (Central American Integration System – SICA)
- **Judith ter Maat**, **Joana Mira Veiga** and **Jos van Gils** (Deltares)
- **Lucilla Minelli** (Food and Agriculture Organization of the United Nations – FAO)
- **Annika Sallisalmi** (Finnish-Swedish Transboundary River Commission – FSTRC)
- **Tassos Krommydas** (Global Water Partnership – Mediterranean – GWP-Med)
- **Peter Johnson** (Great Lakes St. Lawrence Governors & Premiers – GSGP)
- **Lotta Ruokanen** (Helsinki Commission – HELCOM)
- **Susanne Schmeier** (IHE Delft and Scientific and Technical Advisory Panel (STAP) of the Global Environment Facility (GEF)
- **Birgit Vogel**, **Adam Kovacs** (International Commission for the Protection of the Danube River – ICPDR)
- **Isabel Wallnoefer** (International Union for Conservation of Nature – IUCN)
- **Andy Rees** (Plymouth Marine Laboratory – PML)
- **Fanny Tham Ratz** (Race for the Baltic)
- **Agnes Mukami Muriuki**, **Jared Bosire**, **Riccardo Zennaro**, **Avantika Singh** (United Nations Environment Programme – UNEP)
- **Laverne Walker** (United Nations Environment Programme – Cartagena Convention Secretariat – UNEP-CCS)
- **Mohamad Kayyal**, **Alessandro Candeloro** (United Nations Environment Programme – Mediterranean Action Plan – UNEP-MAP)

The Secretariat further extends its gratitude to all those who provided comments throughout the process, in particular to the countries who took the time to review the Guidance Note and provided feedback.

At the Water Convention Secretariat, **Elise Zerrath**, **Julien Favier** and **Sonja Koeppel** led the coordination of this publication's development, with guidance and support from **Francesca Bernardini**. **Lien Mai-Baby** and **Cammile Marcelo** provided dedicated administrative assistance throughout the process.

Finally, the development of this publication was made possible through the generous financial support of Estonia, Italy, Slovenia and Sweden, whose contributions are gratefully acknowledged. The Secretariat also extends sincere appreciation to all Parties and partners who provide financial and in-kind support to the work under the Water Convention.

Every effort has been made to acknowledge all those who contributed. The Secretariat apologises if anyone was unintentionally omitted.

Contents

Contents

List of figures

List of tables

List of boxes

List of case studies

Acronyms and abbreviations

ACTO	Amazon Cooperation Treaty Organization
ADB	Asian Development Bank
BCC	Benguela Current Commission
BSC	Commission on the Protection of the Black Sea Against Pollution
CBD	Convention on Biological Diversity
CCAD	Central American Commission for Environment and Development
CMS	Convention on the Conservation of Migratory Species of Wild Animals
COP	Conference of the Parties
CSO	Civil Society Organization
CSPD/BSR	Committee on Spatial Planning and Development of the Baltic Sea Region
DPSIR	Driver-Pressure-State-Impact-Response
EbA	Ecosystem-based Approach
EFR	Environmental Flow Requirement
EIA	Environmental Impact Assessment
EPA	Environmental Protection Agency
EU	European Union
FAO	Food and Agriculture Organization of the United Nations
FWC	Freshwater Challenge
GBF	Global Biodiversity Framework
GEF	Global Environment Facility
GEF IW:LEARN	GEF International Waters Learning Exchange and Resource Network
GEF-MedProgramme	Global Environment Facility Mediterranean Sea Programme
GIZ	Deutsche Gesellschaft für Internationale Zusammenarbeit (German Agency for International Cooperation)
HELCOM	Baltic Marine Environment Protection Commission (Helsinki Commission)
IADB	Inter-American Development Bank
ICPBS	International Commission for the Protection of the Black Sea
ICPDR	International Commission for the Protection of the Danube River
ICPO	International Commission on the Protection of the Odra
ICPR	International Commission for the Protection of the Rhine
ICWAM	Integrating Watershed and Coastal Areas Management
ICZM	Integrated Coastal Zone Management
IGBP	International Geosphere Biosphere Programme
IHDP	International Human Dimensions Programme
IWRM	Integrated Water Resources Management

LBSA Protocol	Protocol for the Protection of the Mediterranean Sea against Pollution from Land-Based Sources and Activities
LME	Large Marine Ecosystem
MENA	Middle East and North Africa
NAP	National Adaptation Plan
NBS	Nature-Based Solutions
NBSAP	National Biodiversity Strategy and Action Plan
NDC	Nationally Determined Contribution
NGO	Non-Governmental Organization
OECD	Organisation for Economic Co-operation and Development
ORASECOM	Orange-Senqu River Commission
OSPAR	Convention for the Protection of the Marine Environment of the North-East Atlantic
RAP	Regional Action Plan
RBO	River Basin Organization
RSCAP	Regional Seas Conventions and Action Plan
SDG	Sustainable Development Goal
SIDS	Small Island Developing States
SIWI	Stockholm International Water Institute
STAP	Scientific and Technical Advisory Panel (GEF)
UNCCD	United Nations Convention to Combat Desertification
UNCLOS	United Nations Convention on the Law of the Sea
UNDP	United Nations Development Programme
UNECE	United Nations Economic Commission for Europe
UNEP	United Nations Environment Programme
UNEP MAP	United Nations Environment Programme – Mediterranean Action Plan
UNESCAP	United Nations Economic and Social Commission for Asia and the Pacific
UNESCO	United Nations Educational, Scientific and Cultural Organization
UNFCCC	United Nations Framework Convention on Climate Change
UN Water	United Nations Water
WEFE Nexus	Water–Energy–Food–Ecosystem Nexus
WWF	Worldwide Fund for Nature

Main messages

Source-to-sea management is a holistic approach that tackles issues spanning land, freshwater, coastal and marine environments. Actions upstream – such as agriculture, industry or urban development – can have far-reaching effects downstream. For example, pollution or overuse of water in one area can harm fisheries, tourism and communities hundreds of kilometres downstream. Changes in coastal or marine areas can also affect environments and people located upstream. Source-to-sea management builds upon existing management frameworks but broadens their scope, supporting multi-stakeholder and cross-sectoral collaboration to address challenges that cross jurisdictional, geographical and sectoral boundaries.

Benefits of source-to-sea management can span environmental benefits such as restored aquatic habitats or increased carbon sequestration, social benefits such as reductions in waterborne diseases or greater social cohesion, and economic benefits such as increased value of fisheries or a more attractive tourism industry. It can also facilitate peace and security benefits by strengthening regional integration and reducing political tensions through increased coordination on shared objectives.

Transboundary cooperation is pivotal to tackling the complex challenges that span land, freshwater, coastal and marine ecosystems. A robust legal foundation for implementing source-to-sea management in transboundary cooperation already exists: the Water Convention and other global agreements provide a crucial legal framework for implementing source-to-sea management in transboundary cooperation. By bringing together governments, international organizations, non-governmental organizations, businesses, academia and citizens, transboundary cooperation can create lasting, positive impacts across source-to-sea systems.

Unlocking transboundary benefits through source-to-sea management

By linking efforts across sectors and borders, source-to-sea management helps countries coordinate policies, share knowledge and attract new partners. This approach strengthens social, environmental and economic resilience in the face of global challenges, optimizes resource use and supports systemwide sustainable development.

- **Holistic management addresses transboundary challenges:** Source-to-sea management offers a holistic and coordinated approach to addressing challenges across land, freshwater, coastal and marine ecosystems. By integrating social, environmental and economic considerations into decision-making, it helps combat pollution, restore water and sediment flows, reverse biodiversity and habitat loss, and preserve ecosystem services. This approach encourages collaboration across sectors and countries, leading to increased coherence across regional, national and sub-national institutions and instruments with the potential to generate synergistic effects and optimize benefits for the source-to-sea system as a whole. Strengthening transboundary cooperation not only supports healthier ecosystems and communities but also fosters regional stability, conflict prevention and sustainable development.
- **Integrated coordination and policy coherence:** Source-to-sea management unites river basin, aquifer, coastal and marine planning, enabling comprehensive assessment of social, environmental and economic impacts and trade-offs across connected systems. By bridging gaps between freshwater and marine management and aligning sectoral policies and governance levels – including National Adaptation Plans and National Biodiversity Strategic Action Plans – as well as international frameworks, it fosters more effective and sustainable solutions.
- **Strengthened governance and collaboration:** Coordinated efforts among bilateral commissions, basin organizations, and Regional Seas Conventions and Action Plans harmonize upstream and downstream measures. This collaborative approach generates balanced social, environmental and economic benefits across the entire source-to-sea system, while also encouraging regional economic cooperation and investment for long-term gains for both nature and communities.
- **Source-to-sea management brings added value:** It attracts new stakeholders who introduce fresh perspectives, innovative solutions and meaningful incentives for action, particularly from the freshwater and ocean communities. By focusing on systemwide benefits, source-to-sea management optimizes resource allocation and minimizes costs, ensuring a more efficient and impactful approach. Additionally, it supports the development of focused and cost-effective monitoring programmes, strengthening long-term sustainability and informed decision-making.
- **Synergy with other systemic approaches:** Source-to-sea management builds upon Integrated Water Resources Management, Integrated Coastal Zone Management, Ecosystem-based Approaches, Marine Spatial Planning and Water-Energy-Food-Ecosystem Nexus to carry out transformative actions at both the national and transboundary basin levels.

Integrating source-to-sea management into legal and transboundary cooperation frameworks

Integrating source-to-sea management into legal and transboundary frameworks helps countries coordinate efforts to protect shared water, coastal and marine resources, and realize social and economic benefits. International and regional agreements provide the foundation for joint action, ensuring effective solutions across the entire source-to-sea system.

- **Global legal frameworks support source-to-sea cooperation:** Since source-to-sea systems frequently cross national borders, effective management depends on strong transboundary water cooperation among riparian and coastal States. The Water Convention, alongside other multilateral environmental agreements such as the Ramsar Convention and the Convention on Biological Diversity, offers a solid framework for integrating source-to-sea principles into governance and management.
- **Transboundary water agreements enable coordinated action:** Existing transboundary water agreements can serve as a foundation for establishing source-to-sea management in transboundary basins. Basin agreements are important starting points for joint commitments and actions.
- **Regional Seas Conventions and Action Plans facilitate cooperation:** Regional Seas Conventions and Action Plans empower countries to tackle shared environmental challenges while advancing the global mission to protect and sustainably manage marine and coastal ecosystems. By integrating provisions related to catchment and inland waters into Regional Seas Conventions and Action Plans, collaboration with upstream stakeholders is strengthened, enabling a more holistic approach to issues such as water stress, pollution, sediment starvation, saltwater intrusion, overfishing, habitat destruction and biodiversity loss.

Fostering inclusive and coordinated source-to-sea action in transboundary basins

Effective source-to-sea management relies on strong inter-institutional coordination, flexible approaches tailored to local contexts and the active involvement of all stakeholders – from upstream to downstream. By formalizing cooperation, integrating source-to-sea provisions into agreements, and ensuring inclusive participation, countries can create lasting solutions to shared challenges.

- **River, lake and aquifer basin organizations create collaboration opportunities:** Joint bodies such as bilateral commissions and River Basin Organizations play a crucial role in linking inland, coastal and marine stakeholders and fostering collaboration across basin-to-coast systems. Promoting cooperation between basin organizations and Regional Seas Conventions and Action Plans further enhances engagement by bringing upstream and downstream stakeholders into conversation.
- **Strengthening cooperation between River Basin Organizations and Regional Seas Conventions and Action Plans:** Formalizing and operationalizing source-to-sea cooperation between bilateral commissions, basin organizations and Regional Seas Conventions and Action Plans can take multiple forms, including memorandums of understanding, letters of agreement and shared protocols. Strong institutional connections can be reinforced through exchanges, joint expert and working groups, joint projects or programmes, collaborative research and monitoring programmes.
- **Including source-to-sea provisions in new agreements:** New transboundary agreements should explicitly incorporate source-to-sea provisions, where relevant, for example across preambles, objectives, institutional arrangements or implementation sections to ensure a systemic perspective.
- **Integrating source-to-sea perspectives in transboundary cooperation:** Embedding source-to-sea management in transboundary cooperation requires joint bodies to systematically consider social, environmental and economic linkages when developing measures and to ensure coherent monitoring for implementation.
- **Flexible operational pathways:** Source-to-sea management can take shape through various approaches, including joint management plans, coordinated monitoring efforts, engagement between upstream and downstream stakeholders, and shared planning tools. Effective implementation relies on processes tailored to specific contexts, co-designed by riparian and coastal States and communities, and inclusive of key sectors and stakeholders.

- **Address land-based sources of pollution:** Implementing source-to-sea management calls for coordinated strategies to tackle pollution from land-based sources of plastics, fertilizers, wastewater, hazardous chemicals and emerging pollutants. Leveraging global multi-stakeholder partnerships strengthens efforts to reduce pollution at its source, in a cost-effective way, and protect coastal and marine ecosystems.
- **Ensure inclusive participation in decision-making:** It is essential to ensure the active involvement in decision-making processes of often-excluded groups such as ancestral and Indigenous Peoples, marginalized and vulnerable communities, women and youth. Inclusive processes help tailor measures to meet the needs of those most dependent on ecosystem services and ensure that diverse perspectives shape sustainable solutions.

Turning source-to-sea principles into transboundary action

Practical steps guide countries in applying source-to-sea principles to cross border cooperation. By building a shared knowledge base, engaging with a wide range of stakeholders and fostering collaborative action, locally relevant source-to-sea cooperation strategies can be established, including those promoted by the Water Convention.

- **Adaptable implementation framework:** The six-step approach provides a structured way to implement source-to-sea management across varied environments, sectoral contexts, and social and economic impacts. This approach involves:
  - → Step 1: Characterizing the biophysical system;
  - → Step 2: Engaging with key stakeholders;
  - → Step 3: Diagnosing the policy and governance system;
  - → Step 4: Designing the way forward;
  - → Step 5: Co-creating an action plan and fostering an enabling enviornment;
  - → Step 6: Monitoring, evaluation, learning and adaptation.
- **Building a shared transboundary knowledge base:** Establishing a collective foundation of knowledge of the biophysical, socioeconomic and governance context across borders and sectors is essential. Such knowledge can draw from different kinds of information from various sources – existing and new research, models, monitoring data, stakeholder surveys, citizen science, and insights from local communities and Indigenous Peoples – and serve as a guiding reference for setting objectives and enhancing cooperation.
- **Engagement of diverse stakeholders:** Successfully addressing source-to-sea challenges requires the active participation of a broad range of stakeholders, including users of ecosystem services and those from sectors beyond water. Strategic and inclusive engagement is a core principle that ensures solutions are sustainable, equitable and responsive to diverse needs.
- **Governance system analysis:** Assessing both formal and informal institutional arrangements and their management mandates provides critical insights into where improvements are needed and highlights barriers that must be addressed. Addressing governance gaps, overlaps and cross-sector conflicts across the source-to-sea continuum is key to strengthening cooperation.
- **Setting a course:** A shared vision of the social, environmental and economic future condition of the source-to-sea system, collaboratively developed by stakeholders, strengthens the foundation for coherent and effective transboundary source-to-sea management. Identifying the necessary actions and defining the roles of key actors underscores the need for collaborative action. Addressing the conditions needed to bring about a shift away from unsustainable practices reinforces the long-term viability of management strategies.
- **Taking source-to-sea action:** Mainstreaming source-to-sea principles into planning processes, governance frameworks, project design and investment strategies can drive meaningful change in transboundary cooperation. Establishing finance mechanisms and enabling environments that incentivize investment in solutions addressing source-to-sea linkages ensures that social, environmental and economic benefits are distributed across the entire system, while also creating local advantages.

Actionable recommendations to accelerate the adoption of source-to-sea management

Embracing source-to-sea management within transboundary cooperation unlocks powerful, holistic solutions that protect ecosystems, enhance livelihoods and drive global sustainability. Success depends on collaboration across sectors, where governments, non-governmental organizations, financial institutions and private enterprises come together as catalysts for change.

- **Unite for transformative change:** Forge partnerships, align cross-sectoral policies, mobilize system-wide investments, foster innovation and empower local communities to accelerate transformative actions that create lasting impact – ensuring a resilient future for both people and nature.
- **Establish effective coordination mechanisms:** Integrate source-to-sea principles across policies to address upstream-downstream social, environmental and economic impacts at national and regional levels. Establish interministerial and cross-border platforms to harmonize sectors and align governance across jurisdictions to address shared source-to-sea challenges.
- **Engage stakeholders for success:** Empower local governments along rivers and coasts to take active roles in transboundary cooperation, ensuring their concerns are represented in national strategies. Facilitate participatory processes to involve Indigenous Peoples, marginalized groups and local communities in planning and decision-making. Engage freshwater, coastal and marine communities in transboundary cooperation to ensure their concerns are integrated into decision-making processes.
- **Invest in bold, innovative and scalable solutions:** Mobilize dedicated and novel financing to support holistic investments addressing transboundary source-to-sea challenges. Provide long-term funding for ecosystem restoration projects and cooperation between freshwater and marine joint bodies to prevent biodiversity loss, restore ocean health and protect vital ecosystems from further degradation, while also mitigating the growing threats of climate change. Fund community-led projects and capacity-building programmes, contributing to equitable and sustainable outcomes.
- **Catalyse source-to-sea management by strengthening capacity:** Build long-lasting expertise at local, national and regional levels to effectively tackle source-to-sea transboundary challenges. Organize regional forums and communities of practice to share experiences and lessons learned from transboundary source-to-sea initiatives, building bridges between freshwater and marine experts. Provide tools and resources to local communities and marginalized groups to enable them to participate effectively in source-to-sea initiatives.
- **Strengthen the evidence base for source-to-sea management:** Fund studies that explore social, environmental and economic linkages between land, freshwater, coastal and marine ecosystems to strengthen the evidence base for source-to-sea management with a particular focus on transboundary issues. Conduct interdisciplinary research and modelling to enhance understanding of linkages across the source-to-sea continuum to inform policy and management decisions.

Source-to-sea management brings added value to transboundary cooperation by ensuring that the social, environmental and economic implications of geographic and sectoral linkages are included and inform decision-making. Recognizing the deep interconnectedness of land, rivers, lakes, aquifers, coastal ecosystems, seas and the ocean is the first step toward holistic management. These vital water systems transcend national borders, making coordinated efforts among diverse stakeholders essential for protecting ecological health, safeguarding riparian communities, fostering socioeconomic development and ensuring regional stability. Transboundary water cooperation becomes the foundation for effective source-to-sea management, allowing nations to work together to address upstream-downstream and cross-sectoral challenges.

Chapter 1

Introduction

Key Takeaways:

- The objective of this Guidance Note is to enhance understanding of the need for and the benefits of source-to-sea governance in transboundary cooperation.
- The foundations for global support for source-to-sea management can be traced back to the 1972 Stockholm Declaration, the 1992 Rio Declaration on Environment and Development, and more recently to the 2022 and 2025 UN Ocean Conference.
- Existing legal frameworks underscore the need to address the linkages between activities on land and in inland waters, and the resultant impacts on freshwater, coastal and marine ecosystems.
- The Water Convention and other global conventions, regional agreements and treaties can provide the framework for implementing source-to-sea management in transboundary cooperation.
- The Guidance Note can be useful to a wide range of actors in transboundary contexts.

1.1 Why is this Guidance Note needed?

This Guidance Note is intended to support the expansion of transboundary source-to-sea cooperation. It provides an introduction to source-to-sea management, positions it within existing global and regional legal frameworks, presents a six-step process for implementing source-to-sea management in transboundary basins, and suggests key actions that can be taken by a range of actors to accelerate its adoption. Implementing source-to-sea management within the framework of intergovernmental agreements will facilitate integrated decisions that protect ecosystems, improve living conditions and meet the global goals of sustainable development.

Need for source-to-sea management

It is increasingly understood that river basins, coasts and the ocean are interconnected, and that social well-being, ecosystem health and economic prosperity are tied to their collective status. Source-to-sea management addresses the vital linkages between land, freshwater, coastal and marine ecosystems. It focuses on the significant impact that land-based activities have on downstream ecosystems while also addressing impacts that move from downstream to upstream.

However, implementation of source-to-sea management is far from widespread, increasing potential social, economic and environmental risks. Despite certain prominent examples, joint bodies for transboundary water cooperation rarely invite non-riparian coastal States to participate in cooperation.

On the positive side, the International Commission for the Protection of the Rhine (ICPR) cooperates with the Convention for the Protection of the Marine Environment of the North-East Atlantic (OSPAR), and the Baltic Marine Environment Protection Commission (HELCOM) cooperates with the International Commission on the Protection of the Odra (ICPO) on addressing issues that cross the land-freshwater-marine continuum. In addition, the International Commission for the Protection of the Danube River (ICPDR) collaborates with coastal States that are Parties to the Convention on the Protection of the Black Sea and the Black Sea Commission (BSC). Cooperation also exists between the Orange-Senqu River Commission (ORASECOM) and the Benguela Current Convention (BCC). While these examples point to the benefits of source-to-sea transboundary cooperation, their limited number underscore the need to expand these efforts.

Source-to-sea management in the Water Convention

Existing legal frameworks underscore the need to address the linkages between activities on land and in inland waters, and the resultant impacts on freshwater, coastal and marine ecosystems. The Convention on the Protection and Use of Transboundary Watercourses and International Lakes (Water Convention)[1] has among its objectives the protection of the marine environment, including but not limited to coastal areas. This includes protection from land-based sources of impacts. A holistic approach to coordination across sectors and national borders is a cornerstone of transboundary water management. As reflected in the Water Convention, this is supported by and is facilitated through reasonable and equitable utilization of transboundary waters, prevention of transboundary impacts and sustainable management.

The Water Convention and other global conventions, regional agreements and treaties can provide the framework for implementing source-to-sea management within transboundary cooperation. For example, the Water Convention includes relevant source-to-sea provisions such as cooperation to harmonize work on preventing, controlling and reducing transboundary impacts and applying the precautionary and "polluter pays" principles. Source-to-sea management can therefore support the achievement of these aims.

This mutually beneficial relationship can resolve challenges arising from institutional and governance fragmentation through the establishment of joint policies, action plans or vision documents which establish coherence between land, freshwater, coastal and marine management. Their preparation, implementation and enforcement can bring together the relevant national institutions to achieve joint objectives and strengthen coordination across sectors, stakeholders and geographic areas. Existing cross-sectoral coordination at the national level can be foundational for transboundary source-to-sea cooperation. At the transboundary level, source-to-sea aspects can be integrated into river basin management planning, coastal zone management and marine spatial planning.

1 United Nations (1992). *Convention on the Protection and Use of Transboundary Watercourses and International Lakes*. Ch_XXVII_5. Helsinki.

Global support for source-to-sea management

The move towards implementing source-to-sea management in transboundary cooperation is further supported by commitments made in declarations and operational and funding strategies. Notably, the source-to-sea approach was highlighted in the political declaration of the 2nd UN Ocean Conference, "Our ocean, our future, our responsibility".[2] This highlights the importance of source-to-sea approaches in "preventing, reducing and controlling marine pollution of all kinds, from both land- and sea-based sources, including nutrient pollution, untreated wastewater, solid waste discharges, hazardous substances ... through improving our understanding of their sources, pathways and impacts on marine ecosystems", positioning source-to-sea management as a solution to overcoming the challenges to achieving SDG 14. The declaration issued by the 3rd UN Ocean Conference "Our ocean, our future: United for urgent action" (United Nations, 2025) reaffirms the need to adopt a source-to-sea approach "to enhance the resilience of coastal and marine ecosystems and communities". Additionally, the Global Programme of Action (GPA) of the United Nations Environment Programme (UNEP)is an intergovernmental initiative aimed at protecting marine and coastal ecosystems by preventing marine pollution from land-based sources.

1.2 Background

This background section presents the evolution of the source-to-sea concept and growing global support for this approach.

Rio Conventions

The foundations of global support for source-to-sea management can be traced back to the 1972 Stockholm Declaration and, two decades later, to the 1992 Rio Declaration on Environment and Development, where the need to take action to reduce the impacts of human activities on the environment was advanced within the theme of the conservation and management of resources. Chapter 17 of Agenda 21, adopted at the 1992 Rio Earth Summit, calls for nations to anticipate and prevent further degradation of the marine environment and to develop land use practices that reduce run-off of soil and wastes to rivers, and thus to the seas. The three Rio Conventions – the United Nations Framework Convention on Climate Change (UNFCCC), the Convention on Biological Diversity (CBD) and the United Nations Convention to Combat Desertification (UNCCD) – further support the need for coordinated action across the land-freshwater-marine continuum.

Highlighting freshwater-marine linkages

Since 1974, UNEP's regional seas programme has implemented regional frameworks for cooperation, management and protection of shared marine and coastal environments. Concerns about the marine environment have also influenced the provisions of the 1982 United Nations Convention on the Law of the Sea (UNCLOS), Part XII on protection and preservation of the marine environment, including the prevention of pollution from land-based sources (Birnie, Boyle and Redgwell, 2009, p. 382). In 1993, the International Geosphere Biosphere Programme (IGBP) launched a project entitled "Land-Ocean Interaction in the Coastal Zone (Ramesh and others, 2015). The project's Basins Assessment applied a Driver-Pressure-State-Impact-Response (DPSIR) framework globally to delineate status trends and the main anchors of interventions in river basin and lake management. It also assessed pressures on the coastal zone including a study on coastal urbanization and deltas (Crossland and others, 2005). In 2002, the IGBP project collaborated with the International Human Dimensions Programme (IHDP) to enable integration of the human dimensions of global environmental change, across the land to ocean – or source to sea – continuum.

Action Platform for Source-to-Sea Management

Building on these and other relevant foundations, the Action Platform for Source-to-Sea Management (S2S Platform) was launched in 2014 by the Stockholm International Water Institute (SIWI), the United Nations Development Programme (UNDP), the United Nations Environment Programme (including UNEP-DHI and UNEP-GPA), the Stockholm Environment Institute, the Global Water Partnership (GWP) and Delta Alliance. The S2S Platform brings together a network of organizations committed to promoting, implementing and building expertise on source-to-sea management. The S2S Platform has raised awareness on source-to-sea issues, influenced policy dialogues, operational and financing strategies, and brought together the freshwater and ocean communities. It has produced action and policy briefs addressing the need to tackle Sustainable Development Goals (SDGs) 6 and 14 together, and the role of source-to-sea management in climate action and transboundary cooperation from source to sea. Through engagement in global events and processes, it has built support for the holistic management of

2 United Nations General Assembly (2022). Our ocean, our future, our responsibility. A/76/L.72. New York.

land, freshwater, coastal and marine resources that balances benefits for the environment, communities and economies from source to sea (Strategy 2021–2025).

Origins of source-to-sea management

A conceptual framework for source-to-sea governance proposed by Granit and others (2017a) includes a taxonomy of key flows, elements to guide analysis and planning, and a common framework for developing a theory of change. This was followed by elaboration of the conceptual framework and an analysis of select projects from the International Waters portfolio of the Global Environment Facility (GEF) conducted for the Scientific and Technical Advisory Panel (STAP) of the latter (Granit and others, 2017b). The six-step source-to-sea approach presented in the conceptual framework was developed further for application to projects and programmes in a guide for practitioners interested in implementing source-to-sea management (Mathews and others, 2019), and to address the mounting issue of plastic pollution from land-based sources (Mathews and Stretz, 2019). Interest in and implementation of source-to-sea management has been increasing steadily, for example in Scotland (Kellock and others, 2023) and Sweden (Mathews and others, 2023).

1.3 Objective of the Guidance Note

The objective of this Guidance Note is to enhance understanding of the need for and the benefits of source-to-sea governance in transboundary cooperation. The Guidance Note is not legally binding, and does not supersede the legal obligations of the Water Convention. It represents an initial step in building capacity to implement source-to-sea management within transboundary basins. It intends to provide advice on incorporating source-to-sea principles into policies, strategies, plans and programmes. Regional workshops to build capacity in the practical application of source-to-sea management principles presented in the Guidance Note are foreseen within the framework of the Water Convention.

The Guidance Note presents case studies that exemplify the guidance offered. While case studies are used to introduce best practices and share lessons learned, they reflect the currently limited number of examples of source-to-sea management in transboundary contexts. As a result, several case studies are drawn from national rather than transboundary basins or cover activities within the borders of a single nation. Ultimately, the Guidance Note is intended to stimulate collaboration that results in social, environmental and economic co-benefits and catalyses coordinated action across all actors along the source-to-sea continuum.

Origins of the Guidance Note

The Guidance Note resulted from the Global Workshop on Source-to-Sea Management, held in Geneva, 14–16 December 2022, which provided practical learning opportunities on effective source-to-sea management. The workshop was organized under the auspices of the Water Convention serviced by UNECE, under the leadership of Estonia and Slovenia, in cooperation with UN Environment Mediterranean Action Plan (UNEP MAP), the Intergovernmental Oceanographic Commission of the United Nations Educational, Scientific and Cultural Organization (IOC-UNESCO), SIWI, the S2S Platform, GWP, the GEF and its International Waters Learning Exchange and Resource Network (GEF IW:Learn).The workshop identified the need for increased awareness and capacity-building on source-to-sea management. To respond to this need, the Working Group on Integrated Water Resource Management (IWRM) under the Water Convention, at its 2023 meeting, decided to develop the present Guidance Note to increase understanding of the need for and the benefits of source-to-sea management and its significance in IWRM, ecosystem conservation and sustainability.[3] The Guidance Note was developed within the framework of the Water Convention, under the leadership of Estonia, Panama and Slovenia. The process was supported by a lead author, the Water Convention Secretariat, and a drafting group.

3 UNECE (2023). *Key Outcomes of the Global Workshop on Source-to-Sea Management*. ECE/MP.WAT/WG.1/2023/5. Eighteenth meeting, Geneva, 19–21 June 2023.

1.4 Target audience and use of the Guidance Note

Source-to-sea management can be of interest to a wide audience, as it involves multiple stakeholders and sectors due to its cross-border and collaborative nature. Stakeholders operate at different governance levels (subnational, national and intergovernmental) and manage shared water bodies such as rivers, lakes and aquifers, deltas, coasts, nearshore coastal water areas and the ocean. Hence, the Guidance Note can be useful to a wide range of actors in the transboundary context. The Guidance Note can be used by:

- government officials, for example to strengthen cross-sectoral coordination and policy coherence within their borders and in transboundary cooperation and across national targets such as National Biodiversity Strategy and Action Plans (NBSAPs) under the Convention on Biological Diversity, Nationally Determined Contributions (NDCs) and National Adaptation Plans (NAPs) under the Paris Agreement, and at the international level in relation to different relevant international and regional governance frameworks;
- River Basin Organizations (RBO), for example, to integrate coastal and marine issues and extend cooperation to include coastal and marine actors;
- Regional Seas Commissions and Action Plans (RSCAPs), for example, to integrate upstream influences on coastal and marine health and to extend cooperation to include RBOs and upstream actors;
- joint bodies, for example, to extend cooperation across the entire source-to-sea continuum taking into account upstream and downstream linkages;
- financial investors and donors, for example, to incorporate source-to-sea considerations in project identification and assessment and identify projects that address source-to-sea challenges;
- the private sector, for example, to recognize risks from source-to-sea challenges and to identify business opportunities that create benefits for the source-to-sea system as a whole;
- civil society and local communities, for example, to advocate for source-to-sea management, cross-sectoral coordination and upstream-downstream cooperation that creates social, environmental and economic benefits;
- sector practitioners, for example, to support the development and implementation of source-to-sea management through practical application of the source-to-sea approach.

1.5 Scope of the Guidance Note

The Guidance Note consists of the following chapters:

- ***Chapter 2. Understanding source-to-sea management*** introduces core concepts such as the source-to-sea system and key flows. It explains what source-to-sea management is, why it is needed and its benefits. This chapter also covers the contributions of source-to-sea management to sustainable development, climate change, pollution and biodiversity loss, and how it relates to other systemic approaches.
- ***Chapter 3. Source-to-sea management and transboundary cooperation*** provides an overview of contributions that source-to-sea management can make to the objectives of a selection of global legal frameworks and transboundary agreements including through RBOs and RSCAPs, as well as how these frameworks provide support for source-to-sea management. It also shares strategies to establish source-to-sea management.
- ***Chapter 4. Practical steps toward transboundary source-to-sea management*** presents guidance on a step-by-step approach for incorporating source-to-sea management into transboundary cooperation. It covers building a shared knowledge base, designing a way forward, co-creating a collaborative action plan and setting up conditions for its successful implementation. It highlights the strengths, challenges and limitations of source-to-sea management.
- ***Chapter 5. Actionable recommendations for accelerating the adoption of source-to-sea management*** suggests actions that governments, international organizations, non-governmental organizations (NGOs), financial institutions, the private sector, academic and research institutions, and regional and basin organizations can take to advance source-to-sea management.
- ***Chapter 6. Conclusions*** summarizes the main findings.

At a glance: examples of source-to-sea management from around the world

	Title	Region	Focus	Level	Key Partners	Page/ chapter
1	Transformative transboundary water management under climate change with science-informed cooperation in the Rhine River Basin	Europe	Water, sediment	Transboundary	International Commission for the Hydrology of the Rhine Basin, International Commission for Protection of the Rhine, Central Commission for Navigation of the Rhine	p.14 Ch.2
2	Sustainable catchment management through enhanced environmental flow assessment in the Mbarali River catchment, Tanzania	Sub-Saharan Africa	Water, sediment, pollution	National	Sokoine University of Agriculture and National Environmental Management Council, VPO, Ministry of Water, Rufiji Basin Water Board, Ministry of Natural Resources and Tourism, Ministry of Agriculture, PO-RALG, villages, schools, Water Users Association	p.15 Ch.2
3	From research to policy: building understanding of linkages between pollution and ecosystem health in the Moulouya River Basin, Morocco	MENA	Water, biota, pollution, ecosystem services	National	Moulouya River Basin Agency (ABHM)	p.17 Ch.2
4	Adopting source-to-sea management to improve environmental quality in coastal and marine environments in the Western Indian Ocean	Sub-Saharan Africa	Water, sediment, pollution	Transboundary	Nairobi Convention for the Protection, Management and Development of the Marine and Coastal Environment of the Southern and Eastern African Region	p.20 Ch.2
5	Multilevel, multisectoral, and multistakeholder coordination to manage water budgets, levels, and flows in the Great Lakes Basin	North America	Water	Transboundary	Great Lakes St. Lawrence River Basin Sustainable Water Resources Agreement, Great Lakes St. Lawrence River Basin Water Resources Compact	p.23 Ch.2
6	Safeguarding the life cycle of migratory fish in the River Torne and the Baltic Sea with restoration activities and impact assessments in Finland and Sweden	Europe	Biota	Transboundary	Finnish-Swedish Transboundary River Commission	p.25 Ch.2
7	Developing coherence through marine and freshwater scientific collaborations to support source-to-sea management in the Tamar River catchment, United Kingdom	Europe	Pollution	National	Tamar Catchment Partnership (TCP)	p.28 Ch.2
8	Spatial and cross-sectoral integration through source-to-sea management and WEFE Nexus approaches for a Damour Integrated Management Plan, Lebanon	MENA	Water, biota, pollution, ecosystem services	National	Damour Integrated Management Plan, Ministry of Water, National Steering Committee	p.34 Ch.2

	Title	Region	Focus	Level	Key Partners	Page/ chapter
9	Coordination and joint actions for the benefit of water resources, ecosystems and communities along the Sixaola River	Central America	Water, sediment, biota	Transboundary	Agreement on Cooperation for Border Development Panama – Costa Rica, Binational Commission of the Sixaola River Basin, Ministry of Environment of Panama, Ministry of Environment and Energy of Costa Rica, Binational Commission of the Sixaola River Basin	p.46 Ch.3
10	Cooperation between joint bodies on managing nutrients in the Danube River Basin, Europe for local and downstream benefits	Europe	Pollution	Transboundary	International Commission for the Protection of the Danube River, Commission on the Protection of the Black Sea Against Pollution	p.48 Ch.3
11	Cooperation on Target Setting and Pollution Prevention in the Baltic Sea Basin	Europe	Pollution	Transboundary	HELCOM Working Group on Source to Sea Management of Nutrients and Hazardous Substances and Sustainable Agricultural Practices, 2021 Baltic Sea Action Plan	p.52 Ch.3
12	Cooperation on Joint Activities for Improvements from the Lesotho Highlands to the Atlantic Ocean	Sub-Saharan Africa	Water, biota	Transboundary	Orange-Senqu River Commission, Benguela Current Commission	p.54 Ch.3
13	Developing upstream-downstream and cross-country cooperation for sustainable source-to-sea management of the Drin River Basin	Europe	Water, sediment, biota, pollution	Transboundary	Drin Coordinated Action, Memorandum of Understanding for the Management of the Extended Transboundary Drin Basin	p.56 Ch.3
14	Community-based ecological coastal rehabilitation using an ecosystem approach in the Seychelles	Sub-Saharan Africa	Water	National	Waste Enforcement and Permits Division, Environment Department, Ministry of Environment, Energy and Climate Change, MECC, NGOs, communities	p.59 Ch.3
15	Curbing land-based pollution to address marine hypoxia: the GEF-8 Clean and Healthy Ocean Integrated Program (CHO IP)	Global	Pollution	Transboundary	Global Environment Facility, FAO, UNDP, CI, EBRD, NUEP, World Bank, IADB, ADB, CAF	p.61 Ch.3
16	Alliances and coordination with multiple stakeholders for social, environmental, and economic benefits in the Mesoamerican Reef Ecoregion	North & Central America	Stakeholder engagement	Transboundary	Integrated Watershed to Reef Management of the Mesoamerican Reef Ecoregion, National Intersectoral Committees	p.65 Ch.3
17	Regional governance for source-to-sea management in the Tangiers-Tetouan-AlHoceima region, Morocco	MENA	Pollution, ecosystem services	National	Coastal Governance Committee, Directorate for the Environment, TTA Regional Development Plan, Regional Coastal Governance Committee, Regional Steering Committee, Regional Council	p.68 Ch.3
18	Characterizing plastic waste pollution from its source to the sea in Indonesia	Asia	Pollution	National	Indonesian Coordinating Ministry for Maritime Affairs and Investment, > 500 regencies in Indonesia	p.74 Ch.4
19	Narrowing the data and knowledge gap on hazardous chemicals to pave the way for a 'toxic-free' Danube River Basin	Europe	Pollution	Transboundary	Interreg Danube Transnational Programme	p.75 Ch.4
20	Raising stakeholder awareness through a Source-to-Sea Learning Tour of the Orange-Senqu River Basin	Sub-Saharan Africa	Water, sediment, pollution	Transboundary	Orange-Senqu River Commission, Strategic Action Plan, National Action Plans	p.78 Ch.4
21	Building bridges between policymakers, academics, and NGOs for upstream management of nutrient loads and wastewater discharge into the Ganges River and the Bay of Bengal, India	Asia	Pollution	National/ Municipal	Yamuna River, UNEP, Sustainable India Trust, Global Wastewater Initiative, Global Partnership on Nutrient Management, Delhi Development Authority, Ministry of Environment	p.80 Ch.4
22	Community engagement on integrating water, land and ecosystems management in Small Island Developing States	Caribbean	Pollution	Transboundary/ National	UNEP Cartagena Convention Secretariat, Caribbean Public Health Agency	p.87 Ch.4
23	Introducing source-to-sea cooperation in the Amazon River Basin through a Strategic Action Programme	South America	Water, sediment, pollution	Transboundary	Amazon Cooperation Treaty Organization, Specialized Technical Panel, Amazon Network of Water Authorities	p.89 Ch.4
24	Awareness campaign engaging horse owners in Sweden to reduce eutrophication in the Baltic Sea	Europe	Pollution	National/ Transboundary	Federation of Swedish Farmers, Swedish Horse Industry Foundation, Swedish Board of Agriculture, Water Authorities, County Administrative Boards, independent foundations	p.91 Ch.4
25	Improving Mtwapa Creek water quality by use of Constructed Wetland Wastewater Treatment technology in Shimo la Tewa, Kenya	Sub-Saharan Africa	Pollution	National	Kenya Marine and Fisheries Research Institute, Shimo la Tewa Prison (under Kenya Prisons Service), GreenWater Environmental, National Environment Management Authority	p.93 Ch.4
26	Strengthening source-to-sea management in the Mediterranean region	Europe/ MENA	Governance	Transboundary	UNEP/MAP, partner countries	p.100 Ch.4

Chapter 2

Understanding source-to-sea management

Key Takeaways:

- Source-to-sea management is a holistic approach that tackles issues spanning land, freshwater, coastal and marine environments.
- With the objective of enhancing coordination between actors, source-to-sea management provides the environment within which different interests can be balanced.
- Many source-to-sea systems are transnational and/or cross boundaries at other governance levels and transboundary cooperation can accelerate progress on shared commitments providing a common framework for countries to pursue source-to-sea objectives together.
- Implementing source-to-sea management in transboundary cooperation can lead to increased coherence across regional, national, and sub-national institutions and instruments with the potential to generate synergistic effects and optimize benefits for the source-to-sea system as a whole.
- Incorporating source-to-sea perspectives into existing processes will expand understanding of system linkages and be a means for harmonizing the priorities and balancing trade-offs for optimal benefits across the source-to-sea system.

At a glance: case studies in this chapter

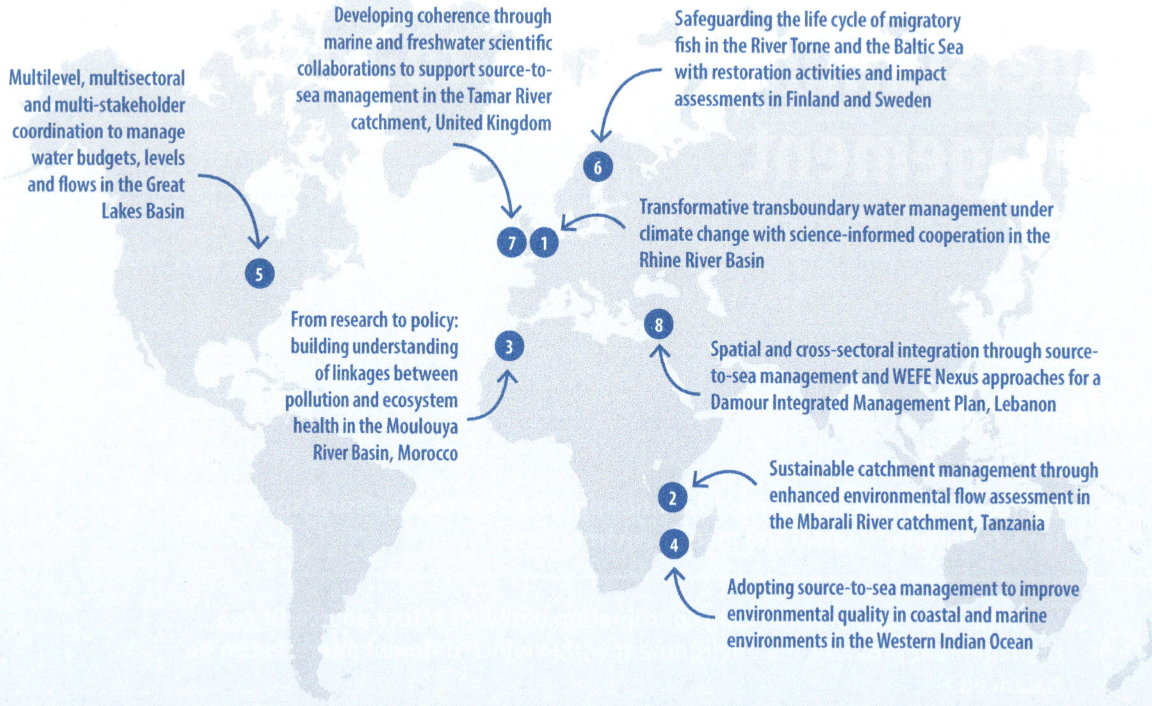

© Gilles Paire / Adobe Stock

2.1 Introduction to source-to-sea management

2.1.1 Source-to-sea systems and key flows

Figure 1 **Source-to-sea system showing the different segments and their interlinkages**

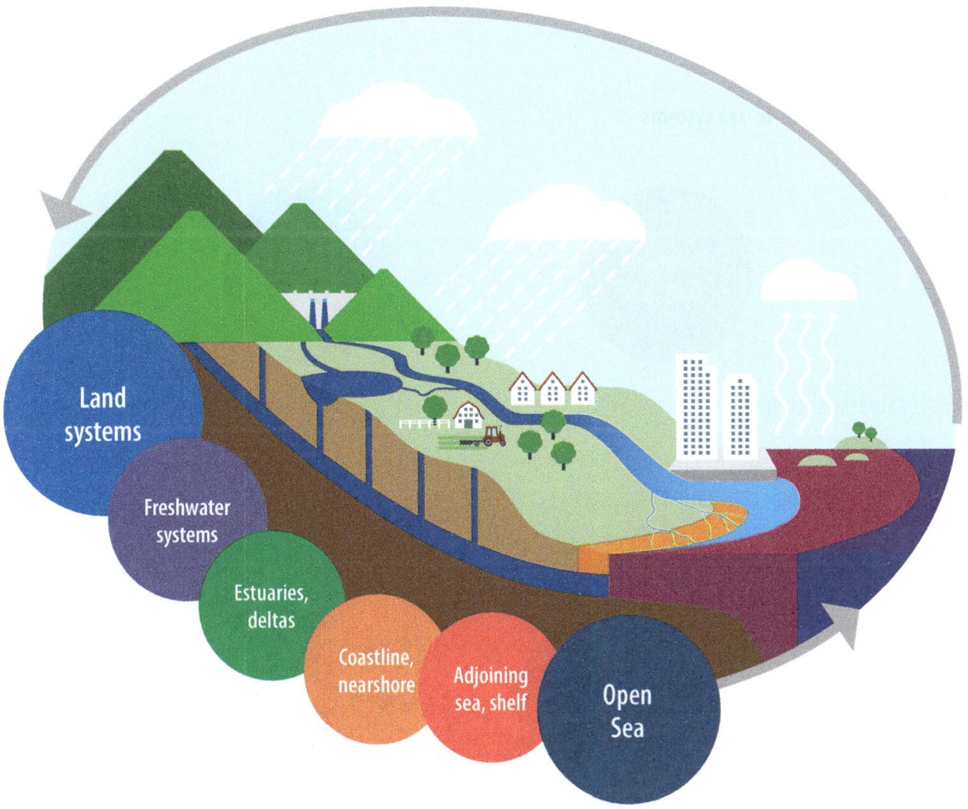

Source: Groeneweg-Thakar, Mathews, Liss Lymer and Sundman, 2020.

The source-to-sea system comprises the land area that is drained by a river system, its lakes and tributaries, connected aquifers and downstream recipients including deltas and estuaries, transitional and near-shore waters, adjoining seas and the continental shelf as well as the ocean (Figure 1) (Granit and others, 2017a). A source-to-sea system can be defined at different scales to reflect the social, economic and environmental objectives of governance. River systems that discharge into closed seas, endorheic lakes or terminal wetlands can also be considered source-to-sea systems. Furthermore, the principles of source-to-sea management can benefit river systems entering lakes and/or reservoirs.

Many source-to-sea systems are transnational and/or cross boundaries at other governance levels. Almost all seas, for instance the Mediterranean, Baltic and Black Seas, and large marine ecosystems, straddle State borders. It is important to note that the social, cultural and economic benefits of ecosystem services can travel from upstream to downstream or the opposite. Examples include glaciers providing downstream water supply and, vice versa, coastal fisheries supplying food and commercial opportunities to upstream communities.

The source-to-sea concept, defined by Granit and others (2017a), defines six key "flows", or biophysical characteristics, that connect the segments of source-to-sea systems – although other "flows" can be of importance. These six key flows are: water, biota, sediment, pollution or water quality, materials or geomorphology, and ecosystem services. The degree of alteration from natural regimes of key flows of water, biota and sediment (blue) and pollutants and materials (red) affect the availability of ecosystem services (green) (Figure 2). In international basins, altered flows in one country can have direct impacts on upstream or downstream neighbours.

Alterations to these key flows can have environmental, social, economic and, subsequently, political consequences, both upstream and/or downstream (Table 1).

Figure 2 Six key flows linking source-to-sea systems

Source: Mathews, Tenberg, Sjödin and Liss Lymer, 2019.

© Nick Fox / Adobe Stock

Table 1 Sources and impacts of alterations to key flows

Water	Changes in land use, water withdrawals for agriculture, industry and household uses, and growing water demands due to population growth and lifestyle changes, coupled with climate change, can contribute to decreased water availability as well as increased flood risk or flood severity downstream. As competition for limited resources grow, water scarcity has been shown to travel downstream as resources are utilized upstream. Changes in the water cycle through human activities can exacerbate water quality issues. While dam and reservoir construction are important measures to manage water flows, mitigate water scarcity and produce low-carbon electricity, the flows of some rivers are so highly diverted by dams and weirs that little water reaches the sea, robbing coastal ecosystems of the water, sediment and nutrients they need. Excessive groundwater abstraction can result in saltwater intrusion to coastal aquifers while nutrients and other forms of pollution that enter groundwater can be transported to coastal and marine waters through groundwater upwellings.
Sediment	Almost a third of global sediment flows are trapped in dams and reservoirs upstream and, without the required flow of sediments from rivers, deltas and coastlines face greater risk of erosion, particularly in the face of sea level rise. In addition, increased sediment loads resulting from activities that cause soil degradation and increased erosion risk in catchments increase turbidity in water bodies, which can reduce primary production and alter aquatic biota populations. When sediment reaches coastal areas, it may smother organisms such as oyster and coral reefs and seagrass beds.
Biota	Populations of riverine, riparian, estuarine and marine organisms can be altered either directly through overharvesting or indirectly through habitat alterations such as changes in water flows, water chemistry and geomorphology. Ecosystem connectivity, both longitudinal (e.g. upstream-downstream) and lateral (floodplain-main channel) is crucial to biodiversity and for maintaining healthy stocks of species that move between different habitats as part of their life cycle. Habitat degradation and loss and blockages of migratory pathways can severely affect these organisms.
Pollution (alterations to water quality)	Untreated or inadequately treated wastewater discharges, agricultural runoff, stormwaters from urban areas as well as solid waste including plastic, marine litter and other waste pose challenges in most regions, and the resulting pollution limits the resilience of freshwater, coastal and marine ecosystems and their ability to perform valuable ecosystem services. Excessive nutrients cause eutrophication, which can alter ecosystems and may impact drinking water supplies. Persistent organic pollutants (POPs), consumer chemicals such as per- and polyfluoroalkyl substances (PFAS), heavy metals and pharmaceuticals have been linked to reproductive, developmental, behavioural, neurological, endocrine and immunological adverse health effects in both humans and wildlife. Plastic, including microplastics, and other solid waste entering waterbodies has seen a rapid increase, affecting environments and human health along the entire source-to-sea continuum.
Materials (alterations to geomorphology)	Construction of infrastructure such as dams, levees and bridges, and other development activities modify river channels and coastlines through, for example, dredging, clearing rocks, extracting of minerals and land reclamation. This can have major implications for water and sediment flows through disconnecting floodplains, reducing aquifer recharge, altering salinity regimes and increasing erosion. Wildlife and plant populations may decrease or disappear entirely due to habitat destruction.
Ecosystem Services	Alterations in one or more of the key flows may result in a reduction of ecosystem services that can span regulating, provisioning, supporting and cultural services. The degradation of freshwater, terrestrial and marine environments directly impacts crucial ecosystem services that provide livelihoods and food security, contribute to the water cycle and water quality, and provide resilience to droughts, floods and climate changes.

Source: Thakar, 2020.

Alterations to one or more key flows become source-to-sea challenges when they cannot be resolved by a single sector, or in a single location, instead requiring collaborative action across sectors, borders or segments of the source-to-sea system. Understanding the alterations to key flows and their significance to social, environmental and economic objectives is essential to support decision-making. For example, to support transboundary cooperation between the Rhine River Basin countries in northern Europe, the International Commission for the Hydrology of the Rhine Basin (CHR) studies three main source-to-sea challenges related to key flows: climate change and hydrology, socioeconomic change and water use, and sediment and morphology (Case Study 1).

1 Case Study

Transformative transboundary water management under climate change with science-informed cooperation in the Rhine River Basin

Context

The Rhine River Basin is one of Europe's major transboundary river basins. The river flows through Switzerland, France, Germany, Luxembourg and the Netherlands, while the basin also stretches into Austria, Liechtenstein, Belgium as well as Italy. The main outlet of the river is the North Sea, although the Rhine also flows into the Wadden Sea (after passing through IJssel Lake). The International Commission for the Hydrology of the Rhine basin (CHR) is the organization through which scientific institutes of the Rhine riparian states develop and disseminate knowledge and tools in the field of hydrology of the entire Rhine catchment area to support decision-making about sustainable development of the basin. As a working alliance of States riparian to the Rhine River, the CHR performs hydrological projections of high and low flows under climate change and impact assessments of socioeconomic developments on the flow regime of the Rhine, among other projects.

Challenge

Glacier melt and changes in snow melt upstream as well as sea level rise downstream will change the hydrological system of the Rhine permanently in the twenty-first century. While more water will be needed downstream for flushing and maintaining (ground)water levels to address saltwater intrusion caused by sea level rise, low flows will become even more extreme, worsening the situation. Water needs to address various socioeconomic demands and increasing climate change will cause additional uncertainty. One example is the projected increase in irrigation for agriculture across the whole basin. Both climate and socioeconomic factors will affect water use and allocation and contribute to extreme water-related events such as floods and droughts. They will also have consequences on water quality, sediment and ecology. Transformative transboundary water management is essential to address these issues.

Activities

The CHR has conducted studies and developed tools to support the development and implementation of adequate mitigation and adaptation strategies. The key organizations involved are the Rhine Commissions – the International Commission for Protection of the Rhine (ICPR) and the Central Commission for Navigation of the Rhine (CCNR) – States, non-governmental orgainizations, knowledge institutions and representatives from various economic sectors. The main sources of funding are governments of the Rhine riparian countries and European funding.

Results

The CHR studies three main source-to-sea challenges in the Rhine River Basin: climate change and hydrology, socioeconomic change and water use, and sediment and morphology. These studies strongly indicate the need for transboundary source-to-sea cooperation. Projections for river discharges, reduced net rainfall and sea level rise under various climate scenarios show how increased glacier and snowmelt in the mountains will increase the length and severity of low flows and increase saltwater intrusion in the delta. The shift to a more rainfed river will introduce more extreme high and low flows and will require better joint operational forecasting and strategic flood and drought management. Growing water demand under various socioeconomic and climate change scenarios will impact water availability and river discharges and, as such, the water supply to different water users, environmental flows and water levels for navigation.

Lessons learned

This case study highlights the growing need for water allocation rules and priorities at the transboundary level for adaptive and integrated water resources management (IWRM) and coastal zone management, while at the same time taking into account specific contexts (e.g. saltwater intrusion, maintaining water levels in peat meadow areas in the delta, melting glaciers and drying up of local groundwater wells in mountainous areas, optimized reservoir management mid-stream for inland shipping, etc.). Addressing sedimentation and riverbed degradation issues upstream to downstream necessitates strong cooperation between (in particular, but not limited to) Austria, Germany and the Netherlands, on sediment monitoring and sediment balance calculation for sustainable sediment management.

Source: Case Study provided by Judith ter Maat, Deltares, 2025.

Environmental Flow Assessments have been conducted to address changes in water flows in the Mbarali River catchment, a key contributor to the Rufiji Basin and, ultimately, coastal and marine ecosystems in Tanzania (Case Study 2). The results were shared in national and regional policy forums, and led to increased recognition of Environmental Flow Assessments as a key tool for IWRM. The Rufiji Basin Water Board (RBWB) pledged to continue supporting conservation efforts, while local governments committed funding to scale up activities. These activities led to increased stakeholder awareness and engagement in catchment conservation.

2 Case Study

Sustainable catchment management through enhanced Environmental Flow Assessments in the Mbarali River catchment, Tanzania

Context

Tanzania's coastal and marine ecosystems are under increasing pressure due to rapid population growth and extensive land-based activities, affecting water quality and quantity. The Mbarali River catchment, a key contributor to the Rufiji Basin, experiences significant degradation from poor land-use practices, overgrazing and deforestation, leading to altered river flows, increased sedimentation and pollution in downstream coastal areas. By addressing land-based sources of degradation, the EFLOWS project contributes to regional efforts under the Nairobi Convention, fostering collaboration between Tanzania and other Western Indian Ocean (WIO) countries to safeguard shared water resources.

Partners included the Sokoine University of Agriculture and National Environmental Management Council, the Vice President's Office (VPO), the Ministry of Water, the Rufiji Basin Water Board (RBWB), the Ministry of Natural Resources and Tourism, the Ministry of Agriculture, the President's Office, Regional Administration and Local Government Tanzania (PO-RALG) , villages, schools and Water User Association (WUA).

Activities

Environmental Flow Assessments were successfully conducted in pilot river sites, producing essential data on hydrological conditions, ecological health and sedimentation trends. Some 45,000 water-friendly tree seedlings were planted to stabilize riverbanks, while WUAs established environmental protection zones with concrete demarcation beacons. The project supported sustainable income-generating activities, including beekeeping and agroforestry, which provided alternative livelihoods and reduced pressure on natural resources. There was initial reluctance among local stakeholders to adopt conservation measures due to competing livelihood priorities. Additionally, limited inter-agency coordination hindered integrated water resources management (IWRM).

Results were shared at national and regional policy forums, leading to increased recognition of Environmental Flow Assessments as a key tool for IWRM. Regional training sessions were conducted for WIO countries, strengthening cross-border collaboration on environmental flow management. The RBWB pledged to continue supporting conservation efforts, while local governments committed funding to scale up activities. In addition, a documentary on environmental flows was produced. All of these activities led to increased stakeholder awareness and engagement in catchment conservation.

Results

The project has increased institutional and human resources capacity at the district and sectoral levels through awareness raising on the importance of environmental flows. The project revealed potential for building a social enterprise or business case particularly at the up-scaling phase. The RBWB is determined to continue supporting the activities initiated by the EFLOWS demonstration project, including the conservation of water sources and the creation of water-friendly tree nurseries.

The local government authorities, particularly the Mbarali District Council, has committed to allocating resources from its annual budget to support WUA activities and to scale up EFLOWS findings. To enhance the income of WUAs, district councils have committed to buy water-friendly and fruit seedlings as part of national tree-planting campaigns. The Resilient Natural Resource Management for Tourism and Growth (REGROW) project has expressed its commitment to enhancing cooperation with WUAs in developing water-friendly tree nurseries and to supporting the management of water sources. They have also provided motorcycles to help WUAs reach the various areas under their jurisdiction.

Lessons learned

Adaptive management is key to successful implementation of environmental flows. Project buy-in is more likely when communities are assured of the direct benefits (e.g. beekeeping, home gardening, etc.). Empowerment of WUAs is critical to the sustainability of restoration actions through nature-based solutions. In addition, Environmental Flow Assessments were found to be an important tool for addressing source-to-sea linkages. The demonstration project provided a foundation for the co-design and co-implementation of nature-based solutions for addressing source-to-sea challenges. Overall, source-to-sea management is fundamental to disentangling siloed planning and enhancing integrated approaches to address linkages along the source-to-sea continuum of land, water, delta, estuary, coast, nearshore and ocean ecosystems. Adopting a landscape/catchment or basin scale is vital to addressing source-to-sea challenges.

Source: Case Study provided by Agnes Mukami Muriuki, UNEP, 2025.

Community members at the tree seedlings, Eflow project in Mbarali River catchment, Mbeya © Winnifred Akiso

In the Moulouya River Basin in Morocco, biodiversity studies improved understanding of the complexity and fragility of the aquatic environment, and the danger of pollutant loads from wastewater, agricultural pesticides and chemical industries. The studies led to investments in wastewater treatment to improve water quality (Case Study 3).

3 Case Study

From research to policy: building understanding of linkages between pollution and ecosystem health in the Moulouya River Basin, Morocco

Context

The Moulouya River Basin, especially its wetlands, is a biodiversity hotspot located in the Mediterranean. The estuary and floodplain complex of the Maghreb's largest Mediterranean river, the Embouchure de la Moulouya, comprises five systems – marine, estuarine, lacustrine, riverine and palustrine, in addition to coastal dunes – and has been designated a Ramsar Wetland of International Significance. The basin is situated in one of the driest areas of Morocco, where recurrent droughts occur. Water supplies are limited due to the arid climate, and the impacts of water scarcity on natural resources are increasing due to the rise in competition for water and pollution from domestic, industrial and agricultural sources. The Protection of Biodiversity and Water Resources in the Moulouya River Basin (PRBREM) project, financed by the European Union, targeted the whole basin, from upstream to downstream. The unifying water resource management body is the Moulouya River Basin Agency (ABHM), which acts as the institutional link with relevant government partners. The studies were conducted by academic partners.

Activities

Key activities included a comprehensive study of the biodiversity of the Moulouya River and its tributaries. This aimed to characterize the type and sources of physicochemical pollution resulting from domestic and industrial wastewaters. Studies carried out by academic partners and experts, appointed as active partners in the PBREM programme, form an essential basis for more rational management of the ecosystem resources of the river basin. Improving skills and developing tools for studying and monitoring biodiversity at the local level are considered crucial to guaranteeing the continuity of obtained results.

Results

After several studies on the quality of drinking water, natural river water and groundwater were initially ignored or even refused, decision-makers eventually accepted the importance of such academic research in the planning and management of regional and national water resources. The project improved understanding of the complexity and fragility of the aquatic environment and of the danger of pollutant loads from wastewater, agricultural pesticides and industrial chemical pollutants. This awareness has contributed to the development of wastewater treatment plants.

Lessons learned

Barriers to source-to-sea management included administrative divisions and the compartmentalization of ministries (water and forestry, agriculture, agencies, offices, etc.), which inhibited coordinated management at the national level. Such issues became even more complex at the cross-border level. One lesson learned from the PBREM programme is that economic, social and ecological considerations must be addressed for source-to-sea management to be successful.

Source: IUCN, 2010.

2.1.2 What is source-to-sea management?

Source-to-sea management is a holistic approach that tackles issues spanning land, freshwater, coastal and marine environments, such as pollution (e.g. nutrients, plastics, chemicals), water-related risks (e.g. floods, droughts, abstraction), sediment (too much or too little), biodiversity (e.g. invasive and migratory species), infrastructure (e.g. dams, levees, canals) and ecosystem services. It achieves this by facilitating increased coordination among sectors and segments of the source-to-sea system and strengthening cooperation across the source-to-sea continuum, with the aim of increasing policy coherence across the source-to-sea system (Mathews and others, 2019).

An assessment of alterations of the six key flows, and their sources and impacts, can function as a starting point for building an understanding of environmental, social and economic linkages across the land-freshwater-marine continuum. This can be developed further by engaging with a range of actors (e.g. policymakers, the private and finance sectors, scientists, NGOs, local communities and Indigenous Peoples) to obtain different viewpoints. To ensure that the provision of critical climate adaptation and ecosystem services by aquatic, riverine, and marine ecosystems are properly considered, these ecosystems should also be represented as stakeholders.

Source-to-sea management aims to enhance coordination between actors, creating an environment in which different interests can be balanced through dialogue and by building a shared vision and common aims for safeguarding ecosystem health. Decision-making is guided by an understanding of upstream-downstream linkages across the entire source-to-sea system, and considers how actions can affect, positively or negatively, other parts of the source-to-sea system. The intended result is to create measurable improvement across the entire source-to-sea continuum.

Dialogue among different actors can facilitate proactive co-creation of individual and joint actions that deliver benefits across the source-to-sea system. This inclusive engagement ensures that trade-offs are carefully examined and that multiple co-benefits are pursued. Co-benefits can be realized when an action by one party leads to gains above and beyond the main benefits intended. Co-benefits also can be achieved when the actions of one party, or in one location, are undertaken in such a way that benefits are accrued by other parties or locations in the source-to-sea system. For example, the installation of a wastewater treatment plant may result in better local water quality and human health, as well as improved downstream water quality resulting in increased coral reef health. Healthier coral reefs can simultaneously provide benefits to coastal communities by reducing shoreline erosion and storm damage. Realizing such benefits which connect upstream and downstream communities is an essential component of source-to-sea management. Improved cross-sectoral coordination can also accrue benefits across goals and objectives. For example, improving coastal habitats through pollution prevention can increase carbon sequestration, conserve biodiversity and contribute to economic development.

Source-to-sea management draws attention to the existing governance system and explores gaps, conflicts and overlaps in governance across the source-to-sea continuum. The assessment can include formal governance measures such as government institutions and instruments as well as informal arrangements. The process provides a baseline for identifying where attention is most needed to improve governance coherence and coverage across the source-to-sea system. In source-to-sea management, inherent biophysical linkages between land, freshwater, coastal and marine ecosystems become the focus for improvement through connectivity, coherence or equivalence in governance.

Various processes can provide the foundation for establishing source-to-sea management. Source: Mathews, Weinberg, Murillo and Liss-Lymer, 2023. presents several examples where source-to-sea management was introduced and key aspects were developed.

Table 2 Overview of the range of processes in which source-to-sea management can be developed

	Bohai-Baltic Seas	**Luga River and Bay**	**KwaZulu-Natal**	**Göta River**	**Three Archipelagos**	**Lake Hawassa Sub-basin**	**Vu Gia-Thu Bon River Basin**
Country/region	Bohai Sea/ Baltic Sea Region (14 countries)	Russian Federation	South Africa	Sweden	Sweden, Finland	Ethiopia	Viet Nam
Ongoing processes to which the study/pilot contributes	Feeds into dialogue through the China Europe Water Platform	Supports the Baltic Sea Action Plan and bilateral cooperation on salmon recovery and reduction of nutrient pollution	Supports the Department of Forestry, Fisheries and Environment in promoting investment by local governments in pollution control	Conducted historical review to identify success factors in non-point source pollution prevention	Analysis of boating and fishing impacts in transboundary archipelago as part of larger project	Contribution to the development of the Strategic IWRM plan for Ethiopian Rift Valley Lakes Basin (2020–2035)	Plastic waste and source-to-sea approach identified as priority by Interprovincial Joint Coordination Committee (Da Nang and Quang Nam provinces)
Size of source-to-sea system	Bohai Sea catchment area (1.3 million km^2); Hai He River Basin area (320,000 km^2); Baltic Sea catchment area (1.7 million km^2); Northern Baltic Sea Water District area (37,000 km^2)	Luga River catchment area (13,600 km^2)	uMngeni catchment area (4,432 km^2); KwaZulu-Natal province area (94,361 km^2)	Göta River catchment area (50,200 km^2)	3 archipelagos within the Baltic Sea catchment area (1.7 million km^2)	Lake Hawassa Sub-basin area (1,400 km^2)	Vu Gia-Thu Bon River Basin area (10,350 km^2)
Priority source-to-sea flows assessed	Plastic waste, nutrient pollution (Nitrogen/ Phosphorus), Hydropower (and its impact on sediment and biota)	Wild Salmon, nutrient pollution (Nitrogen/ Phosphorus)	Solid waste/ plastic waste	Pollutants: Nitrogen, Phosphorus, Mercury Materials: Hydropower and fish	Materials (boating): Biota (fish)	Sediment erosion; plastic waste	Plastic waste

Source: Mathews, Weinberg, Murillo and Liss-Lymer, 2023.

The Convention for the Protection, Management and Development of the Marine and Coastal Environment of the Southern and Eastern African Region (Nairobi Convention) was signed in 1985 and came into force in 1996. The Convention aims to protect and restore the coastal and marine ecosystems in the Eastern African region. A decision at the 11th Conference of Parties of the Nairobi Convention in 2024 urged Contracting Parties to incorporate source-to-sea management into their efforts to control, manage and prevent coastal and marine pollution (Case Study 4). This decision represents the highest level of political commitment by governments of the region to adopt and apply this approach based on sound scientific evidence.

4 Case Study

Adopting source-to-sea management to improve environmental quality in coastal and marine environments in the Western Indian Ocean

Context

The Western Indian Ocean (WIO) region includes 10 countries: Comoros, France (Réunion), Kenya, Madagascar, Mauritius, Mozambique, Seychelles, Somalia, South Africa and Tanzania. The WIO, one of the least ecologically disturbed areas globally, is under threat due to increasing anthropogenic pressures. Over 60 million people inhabit the region and are mostly dependent on coastal and marine resources, which have an estimated value of over $25 billion annually. However, rapid population growth and land-based activities are causing environmental degradation, due to pollution, habitat destruction and changes in freshwater flows and sediment loads. Pollutants flowing into the ocean include household and industrial sewage, solid waste and nutrients. Physical alteration and loss of critical habitats, such as mangroves, seagrass beds and coral reefs, are also major concerns. Climate change exacerbates these threats due to abnormal rainfall patterns, droughts, floods and sea level changes. The countries in the region, many with high poverty levels, need urgent measures to reverse the rapid degradation of coastal and marine ecosystems.

The Nairobi Convention was created by the 10 WIO countries to restore and protect the coastal and marine ecosystems in the region. The Convention is a regional mechanism that enables coordination and collaborative actions between different stakeholders including governments, non-governmental organizations, research institutes and local communities. It has played a pivotal role in improving the sustainable management of coastal and marine ecosystems in the WIO.

Activities

A landmark programme of the Nairobi Convention, designed to advance its goals, is the Strategic Action Programme for the Protection of the Western Indian Ocean from Land-Based Sources and Activities (WIOSAP). This programme is implemented with the support of the Global Environment Facility (GEF) and the United Nations Environment Programme (UNEP). Building on the findings and activities of previous projects, WIOSAP aims to reduce land-based stressors of coastal and marine ecosystems by protecting critical habitats, improving water quality, managing river flows and promoting regional collaboration and governance frameworks that prevent ecosystem degradation.

Challenges

One of the main barriers that delayed the activities of WIOSAP was the different administrative frameworks in each country related to programme activities. These frameworks were sometimes excessively complicated and hindered the efforts of public officials, who acted as focal points leading the implementation of activities in each country. To overcome this barrier, the programme set up national task forces that built a relationship with the focal points and supported them in navigating the different bureaucratic processes required to advance each activity. Additionally, WIOSAP created an informal community of practice by facilitating knowledge-exchange opportunities between focal points from each country. In these sessions, the participants were able to share challenges and solutions they had implemented or discuss alternatives for different contexts.

Another issue faced by WIOSAP was the need to address diverse capacity development requirements across the vast range of stakeholders involved in the programme activities. To address this challenge, WIOSAP mapped the different needs by country and stakeholder, and proceeded to develop tailored programmes that support better implementation of WIOSAP activities.

Results

During the 11th Conference of Parties of the Nairobi Convention, held in August 2024 in Madagascar, informed by the huge investments under the GEF-funded WIOSAP project and the Science to Policy Platform of the Convention, held in December 2023 (also supported by WIOSAP), the Contracting Parties adopted a new Work Programme (2025–2028) and 16 COP Decisions.

Decision CP 11/12 concerns improving environmental quality through source-to-sea approaches. This decision urges Contracting Parties to strengthen their Integrated Water Resource Management (IWRM) frameworks by incorporating source-to-sea approaches and requests the Secretariat in collaboration with partners to support efforts to control, manage and prevent coastal and marine pollution through source-to-sea approaches, and to establish a regional community of practice on IWRM to support the implementation of relevant existing frameworks, processes, guidelines or strategies.

Component 3: Environmental Quality (Adopting the Source-to-Sea (S2S) Approach) of the 2025–2028 Work Programme includes a major focus on source-to-sea management, thereby affirming the highest level of political commitment by governments of the region to adopt and apply this approach based on sound scientific evidence.

Source: Case Study provided by Jared Bosire, UNEP, 2025.

2.1.3 Benefits of implementing source-to-sea management

Incorporating source-to-sea management in transboundary cooperation ensures that the social, environmental and economic implications of geographic and sectoral linkages are included in and inform decision-making. It acknowledges their interconnections and bridges silos to work across sectors and administrative boundaries. Source-to-sea management builds upon existing management frameworks but broadens their scope beyond the common approach of independent management of land, river basins, coasts and the ocean. It supports multi-stakeholder and cross-sectoral collaboration to address challenges that cross jurisdictional, geographical and sectoral boundaries.

Reaching across silos to build concerted action can lead to increased coherence across regional, national and subnational institutions and instruments, with the potential to generate synergistic effects and optimize benefits for the source-to-sea system as a whole. Financial benefits may arise from joint identification and implementation of actions through the avoidance of isolated actions that may negatively impact other parts of the source-to-sea system, or that cannot offer a full remedy to reduce impacts. Through inclusive engagement of stakeholders from different segments and sectors, source-to-sea management can facilitate the formation of action coalitions that catalyse changes in behaviour that will result in greater resilience at the system level. It is complementary to and can be combined with other systemic approaches.

© whitcomberd / Adobe Stock

Benefits can span environmental benefits such as restored aquatic habitats or increased carbon sequestration, social benefits such as reductions in waterborne diseases or greater social cohesion, and economic benefits such as increased value of fisheries due to stronger fish populations or a more attractive tourism industry. There can also be governance benefits including the establishment of clear rules and procedures for joint management, protections for the participation of marginalized stakeholders and the inclusion of the environment as a stakeholder, and improved science-based and cooperative decision-making (Box 1).

Box 1
Benefits of implementing source-to-sea management

- **economic benefits** by addressing the linkages across the source-to-sea continuum in developing specific requirements related to the quality, quantity and timing of water resources for economic activities (e.g. agriculture, industry, energy, nature-based tourism, water-based transport) and reducing the impact of water-related hazards;
- **social and environmental benefits** by improving ecosystems health and providing ecological benefits, as well as social benefits (e.g. health impacts from improved water quality, employment and poverty reduction derived from economic benefits and cultural and recreational benefits);
- **regional economic cooperation benefits** by creating an enabling environment for broader cooperation and investments that create benefits for the source-to-sea system as a whole;
- **peace and security benefits** by including the strengthening of regional integration and mutual dependencies, the reduction of political tensions and the development of dispute resolution tools and approaches with a specific outcome of increased coordination between freshwater and ocean actors through joint bodies and activities;
- **governance benefits** by including the establishment of clear rules and procedures for joint management, protections for marginalized stakeholders and the environment, and improving science-based and cooperative decision-making.

Source: Adapted from UNECE, 2015.

Social, environmental and economic resilience can be created through policy reforms, sustainable financing and by developing frameworks which engender cross-sectoral coordination and facilitate upstream-downstream stakeholder cooperation to reach shared goals and objectives. Coordination across administrative and political boundaries can redress the impacts of longstanding fragmentation of governance. This coordination can be fostered through both formal and informal processes such as ad hoc coordination mechanisms, initiatives or partnerships, and can be supported through existing mechanisms or under new agreements.

In the Great Lakes Basin, North America, the transboundary source-to-sea cooperation enabled by the Great Lakes St. Lawrence River Basin Sustainable Water Resources Agreement and the Great Lakes St. Lawrence River Basin Water Resources Compact has led to stakeholders, governments and the public at large becoming more aware of water use and the importance of data gathering and cooperative and sustainable water use management (Case Study 5).

5 Case Study

Multilevel, multisectoral and multi-stakeholder coordination to manage water budgets, levels and flows in the Great Lakes Basin

Context

The Great Lakes St. Lawrence River Basin in North America ("Great Lakes Basin") is both a series of lakes and an extremely large river system, which empties into the Atlantic Ocean. It has both upstream and downstream interests that must be addressed in any water management regime. Eight US States and two Canadian Provinces have primary management authority over basin water use and work collaboratively with 40 US Federally Recognized Tribes and 144 Canadian First Nations. This authority includes responsibility for water quantity management, water use decision-making, water use data collection, water conservation and efficiency, monitoring and the implementation of water quality laws. They also hold decision-making authority regarding proposals to withdraw and divert water from the Great Lakes Basin. Transboundary cooperation is governed by the Great Lakes St. Lawrence River Basin Sustainable Water Resources Agreement ("Agreement"), and the Great Lakes St. Lawrence River Basin Water Resources Compact ("Compact"), which is a legally binding agreement among the eight Great Lakes States.

Water throughout the Great Lakes Basin is used in a variety of ways. Crucially, it provides drinking water to over 40 million people. It also serves as a shipping corridor enabling the carrying of iron ore from the port of Duluth in Western Lake Superior to the downstream steel mills on Lakes Michigan and Erie, or grains through the Saint Lawrence Seaway to Europe. Basin water is also used for hydropower generation at the downstream end of the system, which in turn relies on water flowing from the headwaters in Lake Superior. In addition, basin water is critical for the continued health of the region's $6 billion fisheries.

The Great Lakes St. Lawrence States and Provinces work together on a range of issues, which include: implementing water management programmes, collecting comprehensive and comparable water use data; implementing water conservation and efficiency programmes; identifying regional scientific research priorities; assessing the cumulative impacts of water withdrawals, diversions and consumptive uses, as well as the anticipated impacts of climate change on the water budget; and engaging in joint decision-making on proposed water diversions of basin-wide interest. To fulfil these commitments, the States and Provinces meet publicly at least twice per year. Funding for this work is provided primarily by the member States and Provinces, with foundations located within the Great Lakes region offering additional support for specific projects.

Multilevel cooperation

The States and Provinces work with multiple levels of government across a transboundary watershed to ensure the development, implementation and execution of multi-stakeholder coordination across the Great Lakes Basin – a complex situation that creates both challenge and opportunities. Working with the proper representatives is important, as well as actively engaging with all key stakeholders – including representatives of industry, agriculture, local government, environmental organizations, conservation organizations, power producers and others. For example, representatives of the Great Lakes Basin Governors and Premiers are charged with making final decisions on the adoption of water use reporting protocols. Throughout the development of these protocols, the representatives engaged with numerous organizations, including:

- an Advisory Committee consisting of approximately 25 regional and national stakeholder organizations;
- a Resource Group of US and Canadian federal agencies, including Environment and Climate Change Canada, the US Environmental Protection Agency, the US Geological Survey, and the National Oceanic and Atmospheric Administration;
- Indigenous organizations including the 40 US federally recognized Tribes in the Great Lake States, the 133 First Nations in Ontario and the 11 First Nations in Québec, with invitations to participate also extended to inter-tribal and First Nation organizations including the Great Lakes Indian Fish & Wildlife Commission and the Chiefs of Ontario.

Involving these people and organizations early in the process allowed for more strategic and thoughtful long-term decisions. This has had a positive and lasting impact on the Great Lakes region.

Results

After the Great Lakes Agreement was signed, all 10 Great Lakes Basin States and Provinces implemented protocols for the collection and dissemination of comprehensive and comparable water use data. The creation and implementation of this system remains key to the success of sustainable water management in the basin. The data have served as a foundation for a series of Cumulative Impact Assessments, produced every five years, which examine the overall water budget for the Great Lakes Basin, as well as for individual lakes (Lakes Superior, Michigan, Huron, Erie and Ontario) and the St. Lawrence River. The Assessments characterize the impacts of withdrawals, diversions and consumptive uses on these water budgets.

In 2023, the Cumulative Impact Assessment also began considering the impacts of climate change, and through academic partnerships, helped to identify the probable impacts of climate change on future Great Lakes Basin water budgets, levels and flows. Such assessments are used by the Great Lakes Basin States and Provinces to individually and together adaptively manage these shared water resources. Ultimately, the goal is to protect and enhance these waters and the Great Lakes St. Lawrence region.

Lessons learned

Treaties and agreements between the US and Canada have existed for a long time. For example, the International Boundary Waters Treaty and the Great Lakes Water Quality Agreement were both in place for years before the more recent water use Agreement and Compact. However, until the Agreement and Compact, the opportunity for cooperative water use management by the Great Lakes States and Provinces – the levels of government with primary water management authority – had not been formalized into law. As a result of the Agreement and Compact, stakeholders, the governments and the public at large are more aware of water use and recognize the importance of data-gathering and cooperative and sustainable water use management. They are also more engaged in the review and consideration of diversion proposals and other higher profile water use activities. The institutions created by the Agreement and Compact provide a useful forum for the States and Provinces to work together on data collection and other forms of information sharing, including best management practices. Engaging in "source-to-sea" coordination and decision-making is required to sustainably manage this shared water resource.

Source: Case Study provided by Peter Johnson, Conference of Great Lakes St. Lawrence Governors & Premiers, 2025.

Involving in decision-making those who are dependent upon ecosystems for their livelihoods, and are most vulnerable to their changes, will contribute local knowledge. This leads to sustainable and enduring outcomes with balanced social, environmental and economic benefits across the source-to-sea system. Raising awareness among upstream stakeholders of their impact on downstream environments, as well as benefits that move upstream from downstream, can strengthen commitments to take action that preserves those benefits.

Through bringing sectors together and including stakeholders across the source-to-sea continuum in decision-making, ecosystem services and their related upstream-downstream benefits can be evaluated alongside the value of economic services (e.g. energy, food production, transportation, etc.). These discussions can highlight trade-offs that can then be assessed against the interlinked goals for sustainable development. Stakeholders can work together to co-create solutions that maintain ecosystem health, while providing social and economic well-being. This then becomes the foundation for collaborative action with stakeholders taking co-responsibility for the implementation of action plans that include both individual and joint actions.

Transboundary cooperation can provide a range of benefits across the source-to-sea system. In the shared waters of the River Torne, bordering Finland and Sweden, transboundary cooperation is leading to the restoration of fish passages between the Baltic Sea and upstream spawning habitats for anadromous salmon and other species. The conservation and recovery of these fish populations implies protecting access to and ensuring the quality of habitats they require along their full life cycle. Cooperation between Finland and Sweden on migratory fish passage and habitats in the river basin enables collaboration on marine spatial planning for wind power (Case Study 6).

6 Case Study

Safeguarding the life cycle of migratory fish in the River Torne and the Baltic Sea with restoration activities and impact assessments in Finland and Sweden

Context

The River Torne, an alpine and sub-alpine watershed shared by Finland and Sweden, is the largest unregulated river system in Europe and home to the most abundant sub-population of the Baltic salmon, thus making it the most significant salmon river in Europe. The value of the recreational fishery exceeds €10 million yearly. Artisanal traditional fishing practices that have been practised for hundreds of years, such as dip-net fishing and seining of salmon and migratory white fish, are still used. Aquatic environments have been altered by clearing tributary channels for timber floating, building roads with culverts, and ditching of forests and mires for drainage. The main channel (i.e. the border river) is unaltered. Recently, the numbers of migrating salmon have decreased dramatically from 50,000 to 100,000 fish to approximately 20,000 in 2023 and 2024.

An earlier agreement for cooperation in the management of shared waters between Finland and Sweden in 1971 was followed by the formation of the Finnish-Swedish Transboundary River Commission (Commission) in 2010. The Commission identifies and takes initiatives in relation to questions of a transboundary character which otherwise might get less attention if managed nationally.

Restoration of the aquatic environment

Funding of €21,364,000 was secured from the European Union for the implementation of the Torne River International Watershed LIFE (TRIWA LIFE) restoration project from 2023 to 2030. The project will restore habitats in the Torne Basin in both Finland and Sweden. The project will tackle challenges salmon face from climate change, such as warming of the Baltic Sea, and fluctuations in herring stocks.

Altogether, 103 hectares of river and stream environments will be restored, 399 obstacles for migratory fish will be removed and 3,366 km of the watershed will be free for migration. Two fish passages will be built in one of the tributaries that has hydropower installations, opening up 970 km of waterways to migratory fish. On top of that, 2,521 hectares of ditched mires will be restored which will have a positive impact on the water balance and water quality in the watershed. Monitoring of results during and after the project is completed is integrated as an essential part of the project.

Salmon is a keystone species for the Torne River and also an important species for fisheries in the Bothnian Bay and beyond in the Baltic Sea. The restoration activities during the TRIWA LIFE project will have a positive effect on the functionality of the ecosystems that form the living environments for migrating salmon and sea trout as well as other aquatic species. Both salmon and sea trout are classified as vulnerable by the Baltic Marine Environment Protection Commission (HELCOM).

In the planning phase of the TRIWA LIFE project, stakeholder contacts and negotiations raised awareness of the positive impact that restoration work would have on tributaries, connectivity and the water balance. The Commission supported the TRIWA LIFE project by engaging with municipalities during its early planning stages. All 10 municipalities in the Torne River Basin agreed to join the project.

Fish migration and energy transition

As migratory fish depend on feeding areas at sea before migrating back to their respective home rivers, concerns have been raised by the Commission about plans for establishing several thousands of offshore wind power installations in the Baltic Sea and their impact on migratory fish. The Commission has asserted that an adequate impact assessment is needed during project permit processes.

The Commission has arranged transboundary conferences to discuss the issue of interactions between off-shore wind power and migratory fish with stakeholders, project owners, researchers and national authorities in charge of marine spatial planning. These meetings have heightened perceptions of the clear need for coordinated assessment of the combined and cumulative effects of offshore wind power on populations of migratory fish, salmon, sea trout and migratory white fish.

Comprehension of the importance of connectivity from both source-to-sea and sea-to-source has grown. The lack of data and knowledge regarding impacts of offshore wind power on migratory fish and aquatic biodiversity has been brought up by the Commission, discussed among stakeholders and has featured in the media. It has become an essential discussion topic in maritime planning processes and, recently, the question has been raised in specific offshore wind power project assessment procedures.

Source: Case Study provided by Annika Sallisalmi, Finnish-Swedish Transboundary River Commission, 2025.

© Finnish-Swedish Transboundary River Commission

2.1.4 Why is source-to-sea management needed?

Governance is often fragmented into silos, organized to respond to societal needs or around components of natural systems, such as river basins or coastal areas. These in turn are administratively divided, for example by national and municipal borders and exclusive economic zones, and separated by sectors (e.g. energy production, agriculture, transportation, urban planning, ports, etc.), all of which are governed by temporal scales much shorter than those of hydro-geomorphologic shifts and climate change. This creates challenges when balancing diverse and often conflicting development objectives, stakeholder priorities, and institutional arrangements that span social, economic and natural borders and timescales.

The ministries and agencies regulating and managing terrestrial, freshwater, coastal and marine issues often differ; institutional mandates may overlap, conflict and/or leave gaps; and there are inconsistencies between terrestrial, freshwater and marine policies and legal agreements and their objectives or targets, such that coordination between them is often difficult. Fragmentation within national boundaries can lead to further complications in the transnational setting with the result of suboptimal outcomes for the source-to-sea system (Granit and others, 2017b).

Governance, moreover, spans a range of functions including policy and strategy, coordination, planning and preparedness, financing, management arrangements, regulation, capacity development, and monitoring, evaluation and learning (Jimenez and others, 2020). In current governance approaches, these functions are undertaken in silos that fail to consider the interactions between different sectors or types of system. For example, discussions around the policy and regulatory measures addressing land-based activities often fail to consider what is needed to achieve sufficient coastal water quality.

In addition, perverse incentives and hidden subsidies may act as underlying drivers that benefit one segment or interest in the source-to-sea system, leading to alterations in key flows that result in impacts elsewhere. Such incentives and subsidies may be driven by financial systems that follow economic borders, often focusing on a specific sector or social geography, and thereby limiting assessments of risks and benefits that could incentivize measures to deliver mutual or co-benefits across the source-to-sea system.

This lack of governance coherence can lead to unintended consequences from unconsidered externalities. Examples include alterations to river flows and sediment transport due to the installation of dams for power generation, irrigation and/or flood control, which in turn affect water quality, lead to habitat loss and starve deltas of sediment replenishment. Practices shaped by this fragmentation of policies, procedures and regulations tend to focus on maximizing local benefits, with limited consideration for upstream or downstream impacts. Furthermore, stakeholder engagement, when it occurs, also focuses on single issues, locations and/or sectors, sometimes competing for the same ecosystem services. This limits the opportunities to increase awareness of linkages across the source-to-sea continuum and identify synergies and co-benefits.

This sectoral fragmentation also has a determining effect on the research conducted, the availability and sharing of data, and the monitoring standards used – all of which play a vital role in informing decision-makers of environmental, social and economic costs and benefits across the source-to-sea system. Accordingly, the interrelationships between activities in one part of the source-to-sea system and the impacts elsewhere are not well understood, and the related scientific evidence is limited.

A key issue is the lack of available integrative management tools such as modelling systems that can facilitate understanding of linkages across land, freshwater, coastal and marine systems. However, in the Tamar Catchment in the United Kingdom, a model that unifies concepts, measures and parameterization processes across freshwater and marine environments has improved understanding of the impacts of land-based activities on coastal and marine environments (Case Study 7). Without the means to understand these relationships, it is difficult to demonstrate the upstream and downstream impacts of proposed governance or financial investments, either positive or negative, in ways that enable appropriate consideration of trade-offs. This results in decisions and practices that are blind to transboundary, cross-sectoral and/or upstream-downstream impacts, and maximize benefits in a single location or sector. The limited availability and lack of coherence in research, data and monitoring mirrors the fragmentation in governance and contributes to its persistence.

7 Case Study

Developing coherence through marine and freshwater scientific collaborations to support source-to-sea management in the Tamar River catchment, United Kingdom

Context

The Tamar River catchment, located in southwest England, covers around 1,800 km^2 and flows into the English Channel. The catchment is populated by an estimated 300,000 inhabitants, the majority of whom are concentrated in the city of Plymouth at the seaward end of the estuary. The most important economic activities in the area are tourism, with around 11 million visitors per year, and agriculture with farmed land covering 73 per cent of the catchment area. The river catchment provides a vast array of ecosystem services including water quality, water availability, greenhouse gas (GHG) sequestration, recreation and cultural spaces, and functional wildlife spaces. However, economic activities in the catchment area are increasing pressure on natural ecosystems and the services they provide. For example, agriculture and animal husbandry substantially increase flows of nutrients and carbon into the water bodies of the Tamar Catchment, including along the coast and into the sea, contributing to issues such as eutrophication and increased release of GHGs.

Activities

The Tamar Catchment Partnership (TCP), coordinated by the Westcountry Rivers Trust, was created in 2011 to guarantee ecosystem services in the Tamar catchment. The TCP is a multi-stakeholder platform formed by landowners, farmers, regulators, recreational users, civil society organizations, national agencies and research institutes, including the Plymouth Marine Laboratory (PML). This platform works to better understand the impacts of human activities in the river catchment, coast and sea, and to manage these activities in a manner that reduces negative impacts.

The Natural Environment Research Council and the Biotechnology and Biological Sciences Research Council of the United Kingdom support the efforts of the TCP by funding projects such as LOCATE and AgZero+. These two projects aim to observe the sources and flows of carbon and nutrients from land to ocean, gather knowledge of GHG sources and sinks, better understand the impacts of agriculture across the source-to-sea continuum and test agricultural practices with reduced impacts. The initiatives brought together several scientific organizations from across the United Kingdom including the Plymouth Marine Laboratory, the National Oceanography Centre, the British Geological Survey, the UK Centre for Ecology and Hydrology, Rothamsted Research and the National Centre for Earth Observation.

Challenges

One of the most significant barriers faced by the project was the need to analyse, in a unified manner, the data gathered from its marine and freshwater research activities, due to differences in the units and methods used. For example, when working on dissolved organic matter in water, freshwater chemists measure the mass (grams per litre), whereas marine chemists use molar quantities (moles per litre). Additionally, the accepted limits of dissolved substances differ between the two. For freshwater science, 0.45 µm is the accepted norm, while for marine science, it is anything that passes through a 0.2 µm filter. These differences in the basic approaches to marine and freshwater science created challenges in information exchange and hampered collaboration between the two communities.

To overcome this barrier, scientists from both the marine and freshwater communities worked together on the LOCATE project and developed the UniDOM model, which unifies concepts, measures and parameterization processes across the freshwater and marine environments. UniDOM allowed the researchers to gain a more complete view of the transport of dissolved material across the source-to-sea continuum by incorporating all ecosystem components. This facilitated a better assessment of how land-based activities impact the catchment, coast and ocean.

Results

The scientific measurements performed by the projects have shown that estuary waters can influence water quality from the coast up to 10 km offshore, and that nutrients and organic material from freshwater bodies contribute to the significant release of GHG from estuarine environments. Furthermore, activities on land to reduce GHG emissions from farmlands, such as applying nitrification inhibitors, may increase the eutrophication of waterbodies. With this information, the TCP can work together with local communities to improve land-based practices that are affecting the functioning of aquatic ecosystems and advance in achieving national and international policy goals such as the UK Marine Strategy, the EU Water Framework Directive, and Sustainable Development Goals 3, 6 and 14.

Lessons learned

The work by PML and its partners in the LOCATE and AgZero+ projects highlight how collaboration between the freshwater and marine scientific communities is crucial to gain a better understanding of linkages along the source-to-sea continuum and the impacts that activities on land have on downstream ecosystems. Collaboration can include practical aspects such as the consolidation of methods and the synchronization of sampling, which increases confidence in the comparability of measurements and their applicability across all environments. Observations made during AgZero+ indicate that nature-based solutions, which have included the creation of wetlands, the introduction of ponds and other approaches to slowing the flow of water, can help in the reduction of nutrients. Going forward, stakeholders will be introduced to possible interventions at the farm and landscape scales to reduce downstream eutrophication.

Additionally, the project narrowed the science–policy gap by carrying out its activities in synergy with the TCP. The new data gathered through the scientific collaboration were made available for the decision-making processes of the TCP, where they were more quickly translated into actions that address the sources of pollution. The combination of science and land management have furthered understanding and improved the ecosystem in this complex and critical environment.

Source: Case Study provided by Andy Rees, Plymouth Marine Laboratory, 2025.

Neglecting social, environmental and economic connections across the source-to-sea continuum can lead to a cascade of impacts that flow both downstream and back upstream. Fragmented governance and mismanaged natural resources can hamper efforts to reach the objectives of global agreements such as the Water Convention, the CBD, the United Nations Convention on the Law of the Sea (UNCLOS),[4] and the UNFCCC Paris Agreement, as well as many regional and transboundary agreements. Global environmental issues such as climate change, biodiversity loss and pollution – as well as growing concerns about water security – provide strong evidence of the need for new governance approaches. Taken together, these failures and the ever-increasing degradation of the natural environment demonstrate the costs of ignoring system linkages.

The COP 29 Declaration on Water for Climate Action[5] underscores

the vital role that protecting, conserving, and restoring water resources, water basins, including seas, rivers and lakes, groundwater and other water-related ecosystems play in delivering effective climate action… highlighting the critical role that water-energy-food-ecosystem nexus plays for both climate mitigation and adaptation, and recognising that water is truly multisectoral and cross-cutting in its nature, and that without addressing water security within all relevant sectors, the climate and sustainable development goals remain out of reach (COP29, 2024).

Source-to-sea management is a critical tool in the efforts to address these complex challenges.

© Riccardo Niels Mayer / Adobe Stock

4 United Nations Convention on the Law of the Sea (1994).

5 COP29, Baku Declaration on Water for Climate Action, 2024.

2.2 Contributions to sustainable development

The 17 United Nations Sustainable Development Goals (SDGs), part of the 2030 Agenda for Sustainable Development, agreed upon by the General Assembly in 2015, form an integrated and indivisible agenda that balances the social, economic and environmental dimensions of development. The interdependencies between the SDGs necessitate coordinated approaches, including source-to-sea management, to achieve the Goals (Figure 3).

Figure 3 **Source-to-sea management for sustainable development**

F Key flows

F.1. Upstream water resources as an example of water as a key flow; **F. 2.** Pollution as key flow, **F. 3.** Hydro power construction as an example of material as a key flow. **F.4.** Sand-dredging with impact on sedimentation as a key flow, **F.5.** Fish as an example of biota as key flow. **F.6.** Ensuring water supply as an example of ecosystem services.

C Challenges and issues

C.1. Upstream water out-take and water treatment, **C.2.** Water out-take/irrigation resulting in lower downstream flows, **C.3.** Increase or decrease in rainfall resulting in risk for downstream flooding or draught, **C.4.** Overexploitation of groundwater aquifers, **C.5.** Construction of infrastructure which impact rivers and their flows. **C.6.** Pollution, including plastics, from industrial production and discharge of wastewater, **C.7.** Nutrients and chemicals from agriculture activities, **C.8.** Energy production providing electricity but potentially blocking upward migration of fish, **C.9.** Loss of biodiversity in water and on land, **C.10.** Reduction of wetlands coverage, **C.11.** Disconnected management of resources and flows within the S2S system with lack of communication and joint target setting between sectors, **C.12.** Lack of transboundary access to resources with implications for migration and conflicts.

M Measures

M.1. Forestry management protecting soils, biodiversity and risk of floods, **M.2.** Connecting rivers to floodplains and aquifers, with biodiversity benefits, **M.3.** Stakeholder consultation between sectors and upstream and downstream stakeholders as a tool for solving mismanagement of key flows, **M.4.** Protecting and restoring reefs for coastal protection and habitat, **M.5.** Conserving and protecting wetlands, mangroves and dunes, **M.6.** Establishing flood bypasses to reduce downstream flooding, **M.7.** Providing riparian buffers to maintain water quality and reduce erosion, **M.8.** Growing crops across slopes to reduce erosion and increase infiltration, **M.9.** Conserving and protecting water sources (i.e. protected areas), **M.10.** Apply a Human Rights Based Approach to safeguard all water-related human rights and that no one is left behind, **M.11.** A special focus on environmental flows in IWRM, **M.12.** Focus on protection and restoration of water-related ecosystems and related ecosystem services, in water governance, for improved human health and poverty alleviation, **M.13.** Strengthened climate change adaptation to improve livelihoods, well-being and resilience, **M.14.** Implement nature-based solutions to improve livelihoods, resilience and human health, **M. 15.** Strengthen inter-sectoral management systems to improve development outcomes. **M.16.** Establishment of inter-sectoral legal and governing frameworks and local government monitoring of compliance.

Source: Sida, 2022.

Greater coordination is needed to ensure that the achievement of one goal does not hinder the advancement of others. With its focus on coordination between sectors leading to coherent governance, source-to-sea management provides a path for coordinated actions to achieve the 2030 Agenda for Sustainable Development as a whole and deliver on the actions agreed upon in the Pact for the Future.[6] Action addressing the whole water cycle can ensure that benefits accrue across its entirety, preserving the integrity of cyclic environmental assets. Source-to-sea management ensures development challenges are tackled holistically and that trade-offs between upstream and downstream stakeholders and across sectors are addressed through open and inclusive dialogues. Additionally, transboundary cooperation can accelerate progress on shared commitments, providing a common framework for countries to pursue source-to-sea objectives together. Pollution, climate change and biodiversity loss have been highlighted as global challenges needing urgent action.

2.2.1 Pollution

The need to control the cumulative effects of multiple sources and the causes of deteriorating water quality in a given water body, from what can be a very large geographical area, represents one of the most cumbersome aspects of water quality management. Pollutants that enter water bodies can be deposited in sediments, bioaccumulated in organisms or transported long distances by rivers and currents, resulting in problems that can be local, regional or even global in nature. The importance of good water quality for both human health and ecosystems is recognized and addressed in the Water Convention and human rights treaties, as well as in explicit terms throughout the SDGs, in relation not only to water and sanitation, but also health, cities, consumption and production patterns, and the ocean.

Addressing the various sources of pollution and their transport along the land-freshwater-marine continuum requires increased understanding of the environmental, social and economic costs of pollution. Moreover, any action must be taken at the pollution source for downstream benefit. Improved understanding of environmental pressures and their potential impacts on the state of the environment is essential to better target responses. Exploring the benefits of pollution reduction in both downstream and upstream areas is an important step in identifying the means for incentivizing action across the source-to-sea continuum. Pollution reduction – the prevention of significant impacts on water quality – is also one of the provisions of the Water Convention.

2.2.2 Climate change

Climate change is one of the defining challenges of the twenty-first century. Climate change calls for holistic governance approaches, due to its interconnections with societal, environmental and economic values, such as water, energy, food and ecological systems. The health of freshwater, coasts and the ocean – from source to sea – is paramount for concerted and ambitious climate action and building resilience. Water, ocean and climate communities have for a long time tackled their areas individually rather than in unison. Source-to-sea management, with its holistic focus on the linkages between ecosystems, is well-positioned to span the boundaries between sectors and scales, and bring these communities together in concerted action on river basin and ocean health, climate mitigation and adaptation.

2.2.3 Biodiversity

Terrestrial, freshwater, estuarine and marine biota have adapted over millennia to available habitat conditions and utilize a range of habitats. Ecosystem connectivity is crucial to biodiversity and for maintaining healthy populations of species that migrate between different habitats as part of their life cycle. Dams and other impediments to movement between habitats risk disrupting these biota flows by reducing connectivity within the source-to-sea system. Habitat loss and degradation, changes in water quantity and quality, or overharvesting in one or more segments of the source-to-sea system can also disrupt biota flows. Addressing biodiversity loss requires better understanding of trade-offs between sectoral development priorities and the benefits derived from biodiversity and ecosystem services. The resolution "Pollution Mitigation by Mainstreaming Biodiversity into Key Sectors UNEP/EA.3/Res.2" calls for mainstreaming biodiversity conservation into key economic sectors such as agriculture, fisheries, infrastructure and tourism, ensuring that pollution prevention efforts also contribute to biodiversity conservation. Building a shared understanding of the benefits from biodiversity conservation through inclusive dialogue with stakeholders across the source-to-sea system will build a foundation for balancing social, economic and environmental objectives.

6 United Nations (2024). Pact for the Future, Global Digital Compact and Declaration on Future Generations. Summit of the Future Outcome Documents.

2.3 Source-to-sea management and other systemic approaches

It is acknowledged that the complexity of the biophysical, social and governance aspects of river basins, coastal zones and marine areas require integrated and systemic approaches. This has led to the development of approaches such as IWRM including the conjunctive management of surface and groundwater, Integrated Coastal Zone Management (ICZM), Ecosystem-based Approaches (EbA), Ridge to Reef, Area-Based Management, Marine Spatial Planning (MSP) and the Water-Energy-Food-Ecosystem (WEFE) Nexus, among others. These approaches are all a means for delineating priorities at different scales and across different sectors. Each has a focal point in which the analysis of interactions is conducted, and priorities are determined.

By focusing analysis and priority setting on the entire source-to-sea system, source-to-sea management can be a means for harmonizing the priorities arising from each of these integrated approaches and balancing trade-offs for optimal benefits. Source-to-sea management creates a holistic view of the values – environmental, social and economic – arising from land, freshwater, coastal and marine ecosystems. By recognizing the interdependencies between them, it strengthens the outcomes from these integrated approaches.

Source-to-sea management builds upon IWRM and ICZM to carry out transformative actions at both national and transboundary basin levels. The IWRM framework takes into account the various users and uses of water, with the aim of promoting positive social, economic and environmental impacts at all levels, including the transboundary level, where appropriate. It sets the basis for undertaking transformative actions by assessing four key components: an enabling environment, institutions and participation, management instruments and financing. It facilitates the cross-sectoral coordination and upstream-downstream cooperation needed for source-to-sea management and enables other enhancing approaches such as WEFE Nexus assessments. Incorporating source-to-sea perspectives into the processes facilitated by these coordination frameworks can lead to an expanded understanding of system linkages and provide opportunities for greater coherence between them (Box 2).

Box 2
Incorporating landscape approaches into source-to-sea management

Management of freshwater and coastal and marine environments remains fragmented, despite increasing recognition of the integrated nature of the water cycle. This has motivated a conceptual study of different systems approaches – including IWRM, ICZM and their integrative, complementary or competitive qualities – and has led to the development of the source-to-sea landscape approach. Analyses of key flows enabled an evaluation of these approaches and the extent to which they consider the water cycle and its interactions with natural and human systems.

The source-to-sea landscape approach was developed to integrate landscapes into source-to-sea management as a structuring element, thereby incorporating hybrid landscape approaches (HLA) and the guidelines from the European Landscape Convention (Michels-Brito, 2023, 2024). The source-to-sea landscape approach proposes a comprehensive view of a territory-based water resource governance and management model that considers the complete water cycle and natural flows that connect different parts of the landscape.

Key takeaways from the source-to-sea landscape study include the following:

- Fragmentation in water and land governance reduces the effectiveness of environmental policies.
- Integrative approaches strengthen sustainability and ecosystem resilience.
- Local actions will only be effective if aligned with a systemic and transboundary view, expanding the scope of analysis to consider the entire continuum from land to ocean, including ecological, social and economic factors.
- Active stakeholder participation in decision-making is essential from the planning stage, including scenario analysis.
- Understanding interdependencies, patterns and processes from past to present is essential.

A variety of integrated approaches have been adopted by the Global Environment Facility Mediterranean Sea Programme (GEF-MedProgramme) to address the complex environmental challenges faced by the Mediterranean region. These include IWRM, conjunctive management of groundwater and surface water resources, the WEFE Nexus, ICZM, ecosystem-based management of marine resources, blue economy development models, climate risk assessment and integration of climate change adaptation measures into development plans and practices. All of these approaches interact with the different, and often multiple, segments of the source-to-sea continuum, defining activity priorities throughout river basins, coastal zones and marine areas. The source-to-sea perspective, therefore, plays an important role in enhanced integration and harmonization of programme and stakeholder efforts for optimal system functioning (Case Study 8).

8 Case Study

Spatial and cross-sectoral integration through source-to-sea management and WEFE Nexus approaches for a Damour Integrated Management Plan, Lebanon

Context

The Damour River Basin, southeast of the Lebanese capital Beirut, represents a critical ecological and agricultural resource. The 40 km perennial river with a 305 km^2 watershed is used for the irrigation of banana and vegetable cultivation along its coastal plains. However, the region faces severe environmental and socioeconomic challenges, such as the consequences of conflicts, displacement, damaged infrastructure, uncontrolled urbanization and inadequate waste management.

Two dams, upstream and downstream, were constructed to divert water for irrigation, but overexploitation by upstream users creates downstream water shortages, negatively impacting downstream agricultural activities, particularly during summer months. This triggers conflicts between the Damour Municipality and upstream communities, which are exacerbated by insufficient, if not non-existent, water allocation legislation. Additionally, the Damour Coastal Aquifer – a primary drinking water source for Beirut – faces seawater intrusion due to overextraction.

Activities

The Damour Integrated Management Plan (IMP), developed under the GEF-funded MedProgramme, and steered by a regional inter-institutional committee led by the National Water Authority, represents the Mediterranean's first IWRM management plan. It combines a source-to-sea approach for spatial integration with the Water-Energy-Food-Ecosystem (WEFE) Nexus assessment and recommendations for intersectoral coordination. This innovative framework addresses water flow and pollution challenges, while ensuring equitable resource access through improved governance and sustainable resources management.

The IMP incorporates several interconnected components: a dedicated Integrated Coastal Zone Management (ICZM) plan, a specialized management plan for the Damour coastal aquifer, arrangements for conjunctive management of surface and groundwater resources, and comprehensive stakeholder engagement across the water continuum. To improve water quality, the IMP includes plans for the upgrading of municipal wastewater treatment plants to provide quality treated water for agriculture and improved marine water quality, linkages with upstream pollution elimination through Polychlorinated Biphenyls (PCB) and mercury removal, and introduction and testing of alternatives to new Persistent Organic Pollutants (POPs). Downstream initiatives include evaluating marine water pollution impacts on fisheries and biodiversity, and the promotion of sustainable fishing practices integrated with biodiversity protection and ecosystem services conservation.

Challenges

Implementation has faced several significant barriers:

- the absence of legal provisions and governance mechanisms for coordinated management across the source-to-sea continuum;
- limited technical capacity among governing authorities to assess water flow requirements;
- weak interconnections between national institutions and with stakeholders for collaborative natural resources management planning;
- ongoing economic crisis and conflicts, exacerbating unsustainable water and waste management practices;
- insufficient integration of IWRM provisions into the national legal framework, undermining their effectiveness;
- limited financial resources enabling pilot actions for sustainable implementation of the IMPs.

Results

The initiative has achieved meaningful progress despite the challenges faced. The Ministry of Water established a National Steering Committee and a coordination mechanism, completed stakeholder and governance mapping, and has developed an initial water resources management knowledge base. This coordination mechanism supports the consultations for elaboration of the IMP at both field and national levels, driving policy coherence, ownership and long-term sustainability. Within the coordination mechanism, improved water governance is addressing conflicts between stakeholders in all segments, focusing on combating resource overexploitation, incorporating climate risk considerations and promoting more equitable cross-sectoral water access.

Lessons learned

Several valuable insights have been gained:

- Source-to-sea management provides an effective framework for evaluating and improving existing water management arrangements and practices.
- The WEFE Nexus and Conjunctive Management of Surface and Ground Water approaches identify water flow issues across different segments and provides sustainable development solutions.
- The Climagine participative design methodology facilitates stakeholder engagement for natural resources management planning including for coherent water rights distribution.

Moving forward, the project will finalize the IMP components, develop an investment plan to support implementation, identify and execute a focused set of pilot actions, drive stakeholder awareness to promote adoption and replication including development of coordination mechanisms, and provide technical support for integrating proposed approaches and tools into national policy and administrative frameworks.

Source: Case Study provided by Mohamad Kayyal and Alessandro Candeloro, UNEP-MAP, 2025.

© Anastasia Rodopoulou

Chapter 3

Source-to-sea management and transboundary water cooperation

Key Takeaways:

- Legal and institutional frameworks for the governance of shared freshwater resources provide valuable starting points for sustainably and effectively governing source-to-sea systems. Frameworks for marine water governance add further important guidance, especially on setting goals related to impacts from land-based activities.

- Global conventions such as the UN Water Convention, the UN Watercourses Convention and the CBD provide general legal principles and obligations facilitating source-to-sea management. Regional conventions or transboundary treaties and agreements translate these general principles into concrete frameworks adapted to the context in the basin or region.

- Linkages between treaties and agreements, RBOs and RSCAPs, whether legal, institutional or related to practical implementation, can provide opportunities to accelerate progress towards source-to-sea management.

- Such cooperation can be facilitated through Joint Task Forces, Memoranda of Understanding, Letters of Agreement and shared Protocols. Institutional arrangements can be implemented according to the context. Examples include integration under a single instrument (e.g. agreements, policies, protocols, MOUs), coordination among different instruments and the creation of joint commissions.

- Advances towards source-to-sea management can be achieved through cooperation between States on the development and implementation of (i) joint action plans; (ii) target setting on pollution prevention; (iii) joint monitoring programmes; (iv) stakeholder engagement; and (v) joint planning instruments.

At a glance: case studies in this chapter

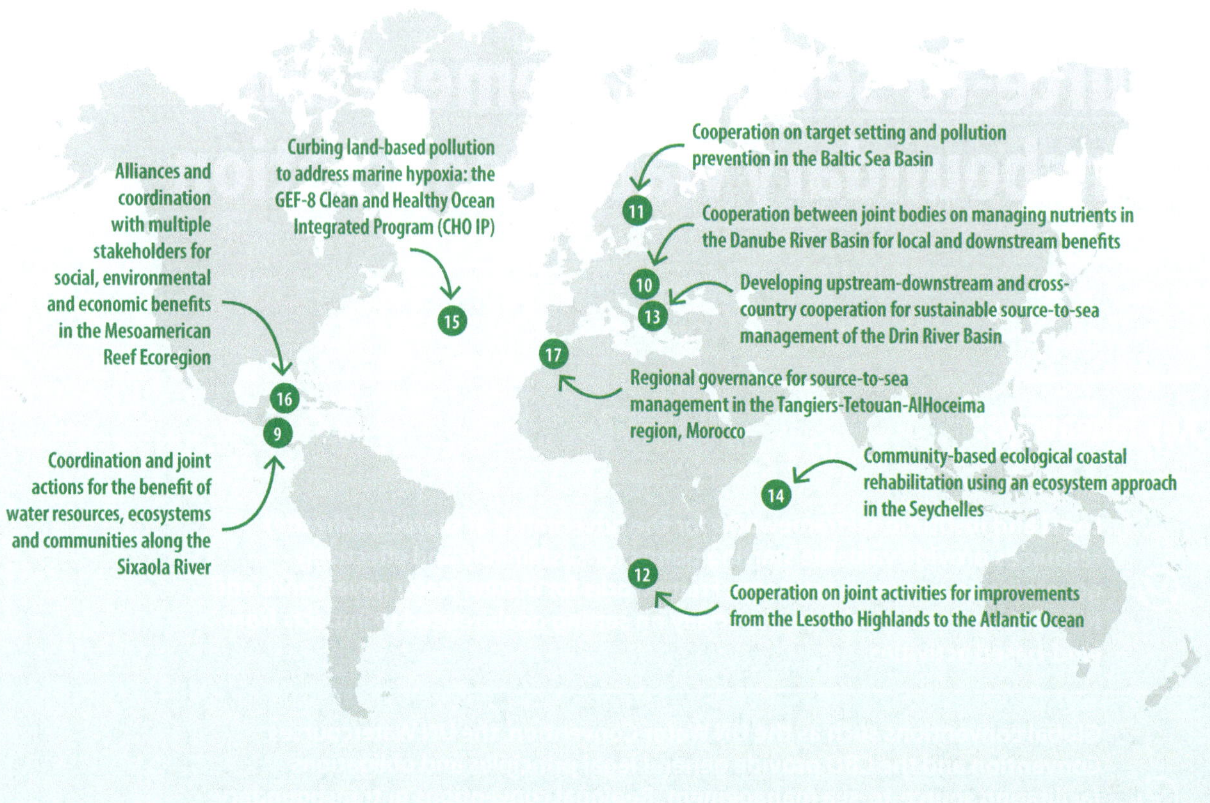

The transboundary nature of many source-to-sea systems presents challenges and opportunities for the integrated, sustainable and effective management of water and related resources along the source-to-sea continuum. The starting point of truly integrated management is to acknowledge that land, rivers, lakes, aquifers, coastal ecosystems, seas and the ocean are interconnected and, thus, require coordinated efforts across multiple stakeholders to maintain or restore ecological health, protect riparian populations, promote socioeconomic development, and ensure peace and stability. As these source-to-sea systems often transcend the boundaries of nation States, transboundary water cooperation between riparian states becomes a prerequisite for effective source-to-sea management.

Over the past decades, States sharing freshwater resources have developed a range of mechanisms and instruments to ensure the cooperative governance of their shared resources. These include international water law principles, the two global water conventions, regional cooperation treaties and institutions, basin agreements and organizations and joint bodies. Together, they provide a comprehensive framework for governing freshwater resources, but so far, they largely lack proper coordination with marine ecosystems and thus integrated source-to-sea governance.

Overall, the existing legal and institutional frameworks for the governance of shared freshwater resources provide valuable starting points for sustainably and effectively governing source-to-sea systems. Frameworks for marine water governance add further important guidance, especially on setting goals related to impacts from land-based activities. However, better integration of these frameworks with a view to increasing policy coherence will be required to ensure that global water resources are managed sustainably and cooperatively for current and future generations.

The legal frameworks that support source-to-sea management can be clustered into two major groups depending on their signatories: global and regional conventions, and treaties or agreements. Unlike global agreements, regional agreements set priorities relevant to their context, which differ from place to place. Global and regional conventions, treaties and agreements are legally binding for the parties that have ratified them. In addition, some global conventions reflect norms of customary international law.

Examples of global conventions include the following:

- 1971 Ramsar Convention on Wetlands
- 1982 United Nations Convention on the Law of the Sea (UNCLOS)
- 1989 Basel Convention[7]
- 1992 Convention on the Protection and Use of Transboundary Watercourses and International Lakes (Water Convention)
- 1992 United Nations Framework Convention on Climate Change (UNFCCC)
- 1992 Convention on Biological Diversity (CBD), including the 2022 Kunming-Montreal Global Biodiversity Framework[8]
- 1994 United Nations Convention to Combat Desertification (UNCCD)
- 1997 Convention on the Law of the Non-navigational Uses of International Watercourses (Watercourses Convention)[9]
- 2001 Stockholm Convention on Persistent Organic Pollutants (POPs).[10]

These legal frameworks may focus on a single important source-to-sea flow or be relevant to one or more source-to-sea segments. For instance, the Stockholm Convention addresses pollutants, with the aim "to protect human health and the environment from persistent organic pollutants" (Article 1, Stockholm Convention), while the Ramsar Convention focuses on "the conservation and wise use of all wetlands through local and national actions and international cooperation, as a contribution towards achieving sustainable development throughout the world."

Examples of regional legal agreements include the following:

- 1974 Convention on the Protection of the Marine Environment of the Baltic Sea Area (Helsinki Convention) (Case Study 11)[11]
- 1976 Convention for the Protection of the Mediterranean Sea Against Pollution (Barcelona Convention) (Case Study 8, Case Study 17)
- 1983 Convention for the Protection and Development of the Marine Environment of the Wider Caribbean Region (Cartagena Convention) (Case Study 22)[12]
- 1994 Danube River Protection Convention (Case Study 10, Case Study 19)
- 1996 Convention for the Development, Protection, Management and Development of the Marine and Coastal Environment of the Western Indian Ocean (Nairobi Convention) (Case Study 4, Case Study 14, Case Study 25)
- 1999 Convention on the Protection of the Rhine (Case Study 1).

Examples of regional, transboundary and bilateral agreements include the following:

- 2010 Agreement between Finland and Sweden Concerning Transboundary Rivers (Case Study 6)
- Agreement on Cooperation for Border Development Panama – Costa Rica (Case Study 9)
- 1978 Amazon Cooperation Treaty (ACTO) (Case Study 23)
- 2008 Great Lakes–St. Lawrence River Basin Water Resources Compact (Case Study 5).

---

7 United Nations Environment Programme (2023a). Basel Convention on the Control of Transboundary Movements of Hazardous Wastes and Their Disposal Protocol on Liability and Compensation for Damage Resulting from Transboundary Movements of Hazardous Wastes and Their Disposal: Texts and Annexes.

8 United Nations Environment Programme (2022). Decision Adopted by the Conference of the Parties to the Convention on Biological Diversity 15/4 Kunming-Montreal Global Biodiversity Framework.

9 United Nations, Convention on the Law of the Non-navigational Uses of International Watercourses (1997).

10 United Nations Environment Programme (2023b). Stockholm Convention on Persistent Organic Pollutants (POPs): Text and Annexes.

11 Convention on the Protection of the Marine Environment of the Baltic Sea Area, 1974, Helsinki

12 Convention for the Protection and Development of the Marine Environment of the Wider Caribbean Region: The Final Act of the Conference of the Plenipotentiaries on the Protection and Development of the Marine Environment of the Wider Caribbean Region Cartagena de Indias, 24 March 1983 Convention on Biological Diversity. Secretariat (2003) Convention on Biological Diversity: Text and Annexes. Montreal.

3.1 Linkages between source-to-sea management and global legal frameworks

3.1.1 Convention on the Protection and Use of Transboundary Watercourses and International Lakes (Water Convention)

The Water Convention covers freshwater systems, estuaries and coastal areas, applying an integrated approach to managing the entire hydrological continuum. Its framework addresses water flows, pollution (e.g. nutrients, hazardous substances), sediment transport and material flows. Adopted in 1992, the preamble of the Water Convention calls for riparian parties to cooperate with coastal States to reduce transboundary impacts on the marine environment, emphasizing the "*need for strengthened national and international measures to prevent, control and reduce the release of hazardous substances into the aquatic environment and to abate eutrophication and acidification, as well as pollution of the marine environment, in particular coastal areas, from land-based sources.*"

The Convention promotes a holistic approach that takes into account the complex interrelationship between the hydrological cycle, land, and flora and fauna, based on the understanding that water resources are an integral part of the ecosystem (UNECE, 2013a). It provides a normative framework for riparian countries to cooperate on collective action necessary for the optimal utilization and protection of their transboundary waters and related ecosystems such as source-to-sea management. The establishment of institutional mechanisms supports permanent and effective cooperation. Parties to the Convention implement individual and cooperative measures to prevent, control and reduce transboundary impacts.

Article 2.2.c of the Convention recognizes the importance of a holistic approach and the need to balance the different interests of stakeholders to allow for the multiple uses of transboundary waters while preserving the ecosystem: "*To ensure that transboundary waters are used in a reasonable and equitable way, taking into particular account their transboundary character, in the case of activities which cause or are likely to cause transboundary impact*".

Article 2.6 states that "*Riparian Parties shall cooperate on the basis of equality and reciprocity, in particular through bilateral and multilateral agreements, in order to develop harmonized policies, programmes and strategies covering the relevant catchment areas, or parts thereof, aimed at the prevention, control and reduction of transboundary impact and aimed at the protection of the environment of transboundary waters or the environment influenced by such waters, including the marine environment.*"

This point is elaborated in Article 9.3: "*In cases where a coastal State is affected by transboundary impact, the Riparian Parties can, if they all so agree, invite that coastal State to be involved in an appropriate manner in the activities of multilateral joint bodies established by Parties riparian to such transboundary waters.*"

© malick / Adobe Stock

Article 9.4 further states that: "*Joint bodies according to this Convention shall invite joint bodies, established by coastal States for the protection of the marine environment directly affected by transboundary impact, to cooperate in order to harmonize their work and to prevent, control and reduce the transboundary impact.*" This facilitates participation by affected coastal States in the activities of the Riparian Parties, with the potential to become a Party to specific transboundary water agreements.

Source-to-sea perspectives should be considered in decision-making processes of basin or river authorities, or of joint bodies managing transboundary waters. The Water Convention stresses the importance of cooperation, either bilateral or multilateral, and the establishment of joint bodies, *inter alias*, to evaluate data to identify pollution sources (Article 9.2.a), to develop concerted action programmes (Article 9.2.f.), and with a view to protecting the marine environment, to cooperate among joint bodies in order to harmonize their work and to prevent, control and reduce transboundary impact (Article 9.4).

Article 13.1 includes the "*obligation to exchange data for all Riparian Parties, whether situated upstream or downstream, for protection of the environment of transboundary waters, as a shared resource, as well as the marine environment.*" It is recommended that non-coastal States located within the catchment areas of transboundary rivers flowing into a regional sea become a Party to such agreements. The Convention defines stakeholders as riparian countries and relevant national authorities. Stakeholder engagement is facilitated through joint bodies or national governments, environmental ministries and local authorities.

3.1.2 Convention on the Law of Non-Navigational Uses of International Watercourses

The Convention on the Law of Non-Navigational Uses of International Watercourses covers freshwater systems and estuaries, taking into consideration protection of the marine environment, and regulates any pollution that leads to "*detrimental alteration in the composition or quality of the waters of an international watercourse which results directly or indirectly from human conduct*" (Article 21). Adopted in 1997, the Convention calls for Watercourse States to take all appropriate measures to avoid causing significant harm, to cooperate for mutual benefit, to exchange data and information, to provide timely notification of and consultation on planned measures, and to protect and preserve the ecosystems of international watercourses (Article 20).

Article 23 stipulates that "*Watercourse States shall, individually and, where appropriate, in cooperation with other States, take all measures with respect to an international watercourse that are necessary to protect and preserve the marine environment, including estuaries, taking into account generally accepted international rules and standards.*" In the event of pollution, this includes setting joint water quality objectives and criteria, establishing techniques and practices to address pollution from point and non-point sources, and establishing lists of substances to be prohibited, limited, investigated and monitored.

Article 24 further states that watercourse States shall enter into consultations concerning the management of an international watercourse. This may include the establishment of a joint management mechanism for "*planning the sustainable development of an international watercourse and providing for the implementation of any plans adopted*" and "*promoting the rational and optimal utilization, protection and control of the watercourse.*"

3.1.3 United Nations Convention on the Law of the Sea (UNCLOS)

The United Nations Convention on the Law of the Sea was adopted in 1982 and entered into force in 1994. Part XII of UNCLOS relating to the protection and preservation of the marine environment addresses the obligation of States to protect and preserve the marine environment. Article 194 indicates that "*States shall take, individually or jointly as appropriate, all measures consistent with this Convention that are necessary to prevent, reduce and control pollution of the marine environment from any source*" and "*take all measures necessary to ensure that activities under their jurisdiction or control are so conducted as not to cause damage by pollution to other States and their environment*" *including minimizing* "*the release of toxic, harmful or noxious substances, especially those which are persistent, from land-based sources.*"

To achieve this objective, Article 197 stipulates that "*States shall cooperate on a global basis and, as appropriate, on a regional basis, directly or through competent international organizations, in formulating and elaborating international rules, standards and recommended practices and procedures consistent with this Convention, for the protection and preservation of the marine environment.*" This may include notification of imminent or actual damage, the formulation of contingency plans, conducting studies and research programmes, the exchange of information and data, and identifying scientific criteria for regulations.

Article 207 obligates States to adopt laws and regulations to prevent, reduce and control pollution from land-based sources defined as rivers, estuaries, pipelines and outfall structures. States "*shall endeavour to establish global and regional rules, standards and recommended practices and procedures to prevent, reduce and control pollution of the marine environment from land-based sources.*" Article 213 further obligates States to enforce their laws and regulations and to take measures in accordance with international rules and standards to prevent, reduce and control pollution of the marine environment from land-based sources.

3.1.4 Convention on Wetlands of International Importance especially as Waterfowl Habitat (Ramsar Convention on Wetlands)

The Ramsar Convention on Wetlands,[13] adopted in 1971, defines wetlands as areas of marsh, fen, peatland or water, whether natural or artificial, permanent or temporary, with water that is static or flowing, fresh, brackish or salty, including areas of marine water the depth of which at low tide does not exceed 6 metres. Wetlands of international importance may incorporate riparian and coastal zones adjacent to wetlands, and islands or bodies of marine water deeper than 6 metres at low tide lying within the wetland. The preamble states that "*the conservation of wetlands and their flora and fauna can be ensured by combining far-sighted national policies with co-ordinated international action*" and that "*waterfowl in their seasonal migrations may transcend frontiers and so should be regarded as an international resource.*"

The Ramsar Convention on Wetlands further states that "*each Contracting Party shall consider its international responsibilities for the conservation, management and wise use of migratory stocks of waterfowl*" and promote the conservation of wetlands and waterfowl by establishing nature reserves (Case Study 2). Article 5 states that "*Contracting Parties shall consult with each other about implementing obligations arising from the Convention especially in the case of a wetland extending over the territories of more than one Contracting Party or where a water system is shared by Contracting Parties. They shall at the same time endeavour to coordinate and support present and future policies and regulations concerning the conservation of wetlands and their flora and fauna.*"

3.1.5 Convention on Biological Diversity and its Kunming-Montreal Global Biodiversity Framework (GBF)

Adopted in 1992, the Convention on Biological Diversity defines biological diversity as "*the variability among living organisms from all sources including, inter alia, terrestrial, marine and other aquatic ecosystems and the ecological complexes of which they are part; this includes diversity within species, between species and of ecosystems.*" Article 5 states that Contracting Parties shall cooperate with each, either "*directly or through competent international organizations, in respect of areas beyond national jurisdiction and on other matters of mutual interest, for the conservation and sustainable use of biological diversity.*" Article 6 further states the conservation and sustainable use of biological diversity should be integrated into relevant sectoral or cross-sectoral plans, programmes and policies. Article 8 establishes that Contracting Parties shall promote the protection of ecosystems, natural habitats and the maintenance of viable populations of species in natural surroundings.

Articles 14, 17 and 18 of the Convention cover the exchange of information between States where activities are likely to "*affect adversely the biological diversity of other States or areas beyond the limits of national jurisdiction, by encouraging the conclusion of bilateral, regional or multilateral arrangements*". States are obliged to "*promote, on the basis of reciprocity, notification, exchange of information and consultation on activities under their jurisdiction or control*", and to facilitate exchange of information and advance international technical and scientific cooperation through international and national institutions in the field of conservation and the sustainable use of biological diversity.

The Preamble of the Kunming-Montreal Global Biodiversity Framework (Box 3), adopted in 2022, invites Parties and other governments to cooperate at the transboundary, regional and international levels in its implementation, while enabling "*participation at all levels of government, with a view to fostering the full and effective contributions of women, youth, indigenous peoples and local communities, civil society organizations, the private and financial sectors, and stakeholders from all other sectors.*" Section A identifies changes in land and sea use and pollution as two of the direct drivers of change in nature with the largest global impact. The purpose of the GBF is to "*catalyze, enable and galvanize urgent and transformative action by Governments, and subnational and local authorities, with the involvement of all of society, to halt and reverse biodiversity loss*". Goal A under the global goals for 2050 consists of ensuring that "*the integrity, connectivity and resilience of all ecosystems are maintained, enhanced, or restored, substantially increasing the area of natural ecosystems.*"

13 Convention of Wetlands of International Importance especially as Waterfowl Habitat, (1971/1982/1987) Ramsar.

Box 3
Global Biodiversity Framework targets relevant to source-to-sea management

Target 1: Ensure that all areas are under participatory, integrated and biodiversity inclusive spatial planning and/or effective management processes addressing land- and sea-use change...

Target 2: Ensure that by 2030 at least 30 per cent of areas of degraded terrestrial, inland water, and marine and coastal ecosystems are under effective restoration, in order to enhance biodiversity and ecosystem functions and services, ecological integrity and connectivity.

Target 3: Ensure and enable that by 2030 at least 30 per cent of terrestrial and inland water areas, and of marine and coastal areas, especially areas of particular importance for biodiversity and ecosystem functions and services, are effectively conserved and managed through ecologically representative, well-connected and equitably governed systems of protected areas and other effective area-based conservation measures, recognizing indigenous and traditional territories, where applicable, and integrated into wider landscapes, seascapes and the ocean...

Target 7: Reduce pollution risks and the negative impact of pollution from all sources by 2030, to levels that are not harmful to biodiversity and ecosystem functions and services, considering cumulative effects, including: (a) by reducing excess nutrients lost to the environment by at least half, including through more efficient nutrient cycling and use; (b) by reducing the overall risk from pesticides and highly hazardous chemicals by at least half, including through integrated pest management, based on science, taking into account food security and livelihoods; and (c) by preventing, reducing, and working towards eliminating plastic pollution.

Target 8: Minimize the impact of climate change and ocean acidification on biodiversity and increase its resilience through mitigation, adaptation, and disaster risk reduction actions, including through nature-based solutions and/or ecosystem-based approaches, while minimizing negative and fostering positive impacts of climate action on biodiversity.

Target 14: Ensure the full integration of biodiversity and its multiple values into policies, regulations, planning and development processes, poverty eradication strategies, strategic environmental assessments, environmental impact assessments and, as appropriate, national accounting, within and across all levels of government and across all sectors, in particular those with significant impacts on biodiversity, progressively aligning all relevant public and private activities, and fiscal and financial flows with the goals and targets of this framework.

3.1.6 Conventions related to pollution across the source-to-sea continuum

Additional global conventions addressing pollution in source-to-sea systems include the Basel Convention (1989), which in Article 1 defines the wastes that are subject to transboundary movement as "hazardous wastes". The Convention regulates transboundary movements in Article 10, which calls for the Parties to cooperate with each other to improve and achieve environmentally sound management of hazardous wastes and other wastes. The Stockholm Convention (2001), sets out, in Article 1, the objective of protecting human health and the environment from persistent organic pollutants. The Convention also lists the chemicals that should be taken into consideration and calls for exchange of information and public awareness and education in Articles 9 and 10, respectively. In the Danube River Basin, a chemical emission model was developed to enhance knowledge and narrow data gaps with regard to monitoring, modelling and management of chemical pollution, and was used to bridge the science-policy interface (Case Study 19).

The Rotterdam Convention (1998)[14] aims (Article 1) to promote shared responsibility and cooperative efforts among Parties in the international trade of certain hazardous chemicals to protect human health and the environment from potential harm and to contribute to their environmentally sound use. The Minamata Convention (2013) aims to protect human health and the environment from anthropogenic emissions and releases of mercury and mercury compounds. The Convention calls for information exchange and public information, awareness and education in Articles 17 and 18, respectively. As of 2025, an international legally binding instrument on plastic pollution, including in the marine environment, is under negotiation.

14 United Nations Environment Programme & Food and Agriculture Organization of the United Nations (2023). Rotterdam Convention on the Prior Informed Consent Procedure for Certain Hazardous Chemicals and Pesticides in International Trade: Text and Annexes.

3.1.7 Conventions related to migratory fish across the source-to-sea continuum

The Convention on the Conservation of Migratory Species of Wild Animals (CMS), adopted in 1979 and entered into force in 1983, is an international treaty that aims to protect migratory species and their habitats. It provides a global platform for coordinated conservation efforts among countries where these species travel, known as Range States. The Convention focuses on species that migrate across national borders and are at risk due to habitat loss, climate change and human activities. The treaty classifies species into two appendices:

- Appendix I: Species that are endangered and require strict protection
- Appendix II: Species that would benefit from international cooperation.

The Convention encourages countries to create agreements tailored to specific species or regions, ranging from legally binding treaties to informal memoranda of understanding. It complements other conservation efforts and works alongside organizations, NGOs and governments to safeguard migratory wildlife.

At a regional level, the Convention for the Conservation of Salmon in the North Atlantic Ocean was adopted in 1982 to protect wild Atlantic salmon populations that migrate beyond national fisheries jurisdictions. It established the North Atlantic Salmon Conservation Organization (NASCO), which coordinates conservation efforts among member countries. The Convention prohibits salmon fishing beyond 12 nautical miles from the coast, except in designated areas like West Greenland (40 nautical miles) and the Faroe Islands. It promotes scientific research, stock restoration and sustainable management practices to ensure the long-term survival of Atlantic salmon.

Also at a regional level, the Convention for the Conservation of Anadromous Stocks in the North Pacific Ocean was signed on 11 February 1992, and entered into force on 16 February 1993. It established the North Pacific Anadromous Fish Commission (NPAFC), which promotes international cooperation to conserve anadromous fish stocks, such as Pacific salmon and steelhead trout, in the North Pacific Ocean. The Convention prohibits directed fishing for anadromous species in international waters beyond the 200-mile exclusive economic zones of coastal states. Member countries – Canada, Japan, the Republic of Korea, the Russian Federation and the United States – collaborate on scientific research, enforcement measures and conservation strategies to protect these migratory fish populations.

3.2 Linkages between source-to-sea management and regional governance frameworks

Basin agreements and Regional Seas Conventions and Action Plans translate global commitments into mechanisms for cooperation in shared basins and seas.

3.2.1 Transboundary cooperation in river basins

3.2.1.1 Basin agreements

The Water Convention provides a global legal framework for transboundary cooperation on source-to-sea management, allowing riparian parties to invite coastal States to become involved, in an appropriate manner, in the activities of joint bodies established by the riparian parties. This engagement aims to reduce transboundary impacts on the marine environment. It further emphasizes the need for cooperation between riparian States on water quality (e.g. nutrients, hazardous substances), sediment flows, pollution prevention and ecosystem conservation.

Existing transboundary water agreements developed by Parties to the Water Convention, as well as by non-Parties, can serve as a means for establishing source-to-sea management in transboundary basins. Basin agreements are often important starting points for cooperative and integrated approaches, as they engage riparian States in cooperation over shared water resources, operationalize more general international water law principles and institutionalize cooperation. Accordingly, they form the basis of riparian States' joint commitments and joint actions in a basin, and can also constitute the starting point for joint actions among States along the source-to-sea continuum.

Basin agreements can address one or more of the key source-to-sea flows and/or specific characteristics of source-to-sea management (e.g. cross-sectoral coordination, stakeholder participation, or monitoring and assessment). Some basin agreements already include explicit commitments that relate to source-to-sea management. For instance, the 1999 Rhine Convention articulates contracting parties' commitment to "help restore the North Sea" (Article 3 (5)). Many measures undertaken in the context of the Convention have indeed contributed to reducing the pollution loads of the Rhine, benefiting the North Sea as the recipient marine ecosystem.

© Damian / Adobe Stock

Likewise, the Preamble of the 1991 Convention on the International Commission for the Protection of the Elbe expresses the need to decrease the pollution load of the North Sea. It includes a commitment for contracting parties to "reduce substantially the pollution of the North Sea from the Elbe area" (Article 1 (2)(c)). Similarly, one of the goals of the Convention on the International Commission for the Protection of the Oder, set out in Article 1, paragraph 2 (a), is "to prevent the pollution of the Oder and the Baltic Sea by contaminants and to achieve a sustained reduction in the pollution thereof".

The Danube River Protection Convention in its preamble emphasizes "the urgent need for strengthened domestic and international measures to prevent, control and reduce significant adverse transboundary impact from the release of hazardous substances and of nutrients into the aquatic environment within the Danube Basin with due attention also given to the Black Sea". It further notes that the contracting Parties are "striving at a lasting improvement and protection of the Danube River and of the waters within its catchment area in particular in the transboundary context and at sustainable water management taking duly into account the interests of the Danube States in the field of water use and at the same time contributing to the protection of the marine environment of the Black Sea". Article 2, paragraph 1 elaborates on this point, stating that contracting Parties "shall endeavour to contribute to reducing the pollution loads of the Black Sea from sources in the catchment area".

However, most basin treaties do not include any reference to the recipient marine water body (Programme in Water Conflict and Transformation, 2025a). This limits the reach of these treaties to the freshwater part of the source-to-sea system. Accordingly, source-to-sea commitments, objectives and cooperation mechanisms could be more explicitly incorporated into many more basin treaties. This could include:

- integrating commitments to sustainably managing or protecting recipient marine waterbodies in the objectives of treaty-based cooperation;
- explicit establishment of coordination mechanisms with marine governance treaties and institutions in the governance arrangements established by a treaty;
- specific governance provisions, such as obligations to engage with actors representing the marine sector or to monitor impacts on marine recipient water bodies.

Where new transboundary agreements are being developed, they should include provisions for cooperation on source-to-sea management, where relevant. Source-to-sea management can be included in the core building blocks of transboundary agreements, such as the Preamble, Objectives and Scope, and in the focus of Joint Bodies. Its inclusion is also relevant to the Principles, Procedural and Implementation sections, Sectoral and Intersectoral Issues, and sections about Water Management and Protection (UNECE, 2021).

3.2.1.2 Institutions for cooperation

The pursuit and achievement of water management objectives often depend on institutionalized cooperation mechanisms that extend beyond the mere existence of treaties, to provide means for developing and implementing joint action. Cooperation is more effective when dedicated institutions, such as joint bodies and regional seas commissions, are created and work together. From the institutional perspective, States sharing transboundary waters can consider several models to address impacts to marine ecosystems coming from activities in river basins.

Engagement between States is facilitated through joint bodies such as RBOs and Basin Commissions, formed to operationalize transboundary agreements. Joint bodies or basin organizations fulfil these activities in over 100 of the world's 313 transboundary basins (Programme in Water Conflict and Transformation, 2025b). They can ensure regular meetings, organize data and information exchange as well as joint monitoring, and implement joint activities or development projects. They also provide important decision-making and dispute-resolution support. RBOs and Basin Commissions thus are an essential institutional element of the bilateral or multilateral regulatory framework of cooperation on transboundary waters.

To promote transboundary cooperation in water resources management, improve the health of shared ecosystems and ensure equitable access to water for communities in the basin, Costa Rica and Panama have established the Binational Commission of the Sixaola River Basin. The two countries have cooperated on the shared project "Towards Integrated Management of Transboundary Water Resources (IWRM) of the Binational Sixaola River Basin". This key initiative strengthens coordination and joint actions for the benefit of water resources, ecosystems and communities dependent on this basin and its coastal and marine ecosystems (Case Study 9).

9 Case Study

Coordination and joint actions for the benefit of water resources, ecosystems and communities along the Sixaola River

Context

The Sixaola River, found on the border between Costa Rica and Panama, is of great ecological importance and crucial for both nations in terms of water resources and biodiversity. The river is also essential for local communities and marine ecosystems, which depend on the quality of water and sediments from the basin. Formal transboundary agreements, such as the Agreement on Cooperation for Border Development Panama – Costa Rica and the Binational Commission of the Sixaola River Basin (CBCRS), facilitate cooperation and coordination between both countries in the management of water resources in the basin. In addition, the relevance of regional agreements on the conservation of biodiversity and marine resources is recognized, reinforcing the need for a shared approach to basin management. Priority environmental problems affecting the river basin include water pollution, sedimentation in water sources and loss of biodiversity.

The project "Towards Integrated Management of Transboundary Water Resources (IWRM) of the Binational Sixaola River Basin", shared between Costa Rica and Panama, also known as the "Connecting Communities and Ecosystems Project", is a key initiative that strengthens coordination and joint actions for the benefit of water resources, ecosystems and communities dependent on this basin. Implementation is coordinated by the Ministry of Environment of Panama (MiAMBIENTE) and the Ministry of Environment and Energy of Costa Rica (MINAE), together with the Binational Commission of the Sixaola River Basin. Local and international non-governmental organizations are also involved in this process, supporting awareness-raising and conservation activities. Local communities, farmers, fishers and other key actors in the basin are the main beneficiaries of sustainable management of water resources and ecosystems, and are vital stakeholders in the success of the project.

Activities

The objectives of this project are to promote transboundary cooperation in water resources management, improve the health of shared ecosystems and ensure equitable access to water for communities in the basin. The main activities include the collection of hydrographic and water quality data in the binational basin, the organization of workshops and roundtables with local communities and key stakeholders, and capacity building in IWRM in both countries. In addition, awareness-raising campaigns on the importance of biodiversity conservation in the basin have been carried out, alongside the development of joint policy and strategy proposals.

Financing for project activities comes from funds provided by multilateral organizations including the Global Environment Facility (GEF) and the Inter-American Development Bank (IDB). At the local level, municipal governments and communities have allocated funds for local initiatives.

Results

Project activities have significantly raised awareness of the interdependence of ecosystems in the basin, both upstream and downstream. The project strengthened local capacities through workshops and specific training. Through Causal Chain and Issue Prioritization workshops, participants gained an understanding of how management practices in one part of the basin can affect communities and ecosystems in other areas, especially in relation to water flow, sedimentation and biodiversity.

The impacts of the project include improved shared management of water resources and increased awareness of the importance of involving women's groups in Indigenous and Afro communities in management of the basin's ecosystems.

The Binational Commission of the Sixaola River Basin remains a platform for open dialogue between all key sectors and actors in the basin. Its principal function is to ensure the integrated management of the Sixaola River hydrographic basin, evaluating the contributions of all projects executed in the basin.

Lessons learned

The Binational Commission of the Sixaola River is an excellent example of integrated management of transboundary basins under the "source-to-sea" approach, integrating actions for the conservation and management of water resources from the upper areas of the basin to the sea, and allowing for a collaborative response at a binational level, guaranteeing the sustainability of the basin and its biodiversity.

A key lesson was the importance of continuous prior consultation on the part of the project with the Binational Commission, which ensured their engagement in all aspects of project activities. Lessons learned include the importance of ensuring the presence of effective project communication mechanisms between both countries, based on existing agreements.

Source: Case Study provided by Karima Lince, Ministry of Environment, Panama, 2025.

The benefits of transboundary source-to-sea cooperation do not flow in only one direction – from upstream to downstream; they can also flow upwards from downstream areas resulting in upstream benefits. Sharing information on upstream benefits from actions taken for the protection of the marine environment can be used to stimulate cooperation. This can be an important starting point for integrated governance along the source-to-sea continuum.

Fostering cooperation where institutions for both freshwater and marine management exist can be a critical step towards integrated transboundary source-to-sea management. In most source-to-sea systems, institutionalized cooperation between institutions governing freshwater basins and institutions governing marine ecosystems remains, however, non-existent.

Various opportunities exist to enhance such cooperation to implement effective source-to-sea management (Case Study 10). These include the establishment of a regular meeting routine, developing integrated source-to-sea management plans that extend beyond traditional basin management plans and include recipient marine areas and their challenges and needs, or joint monitoring based on mechanisms and indicators that incorporate the needs of marine ecosystems. Institution building might be required in basins and marine areas where no institutions exist or where institutions only span parts of the source-to-sea system.

10 Case Study

Cooperation between joint bodies on managing nutrients in the Danube River Basin for local and downstream benefits

Context

The Danube River Basin (DRB) is shared by 19 countries and connects the Black Forest in Germany to the Black Sea northwest shelf shared by Bulgaria and Romania. It drains into the Black Sea, which is mostly hydrologically isolated and has experienced ecological problems due to fluxes of pollutants (nutrients, hazardous substances) from the Danube River, among other sources. Since 1994, the International Commission for the Protection of the Danube River (ICPDR) has been coordinating transboundary water management in the DRB and facilitating the implementation of the EU Water Framework and Floods Directives. The Commission on the Protection of the Black Sea Against Pollution (BSC) is a strategic partner of the ICPDR, safeguarding the Black Sea marine environment in line with the EU Marine Strategy Framework Directive.

Since measures applied in the DRB to reduce pollution of water bodies also contribute to the improvement in the status of the Black Sea, the joint implementation of source-to-sea management is a key objective of collaboration between the two Commissions.

Activities

To safeguard the status of the Black Sea ecosystems, the ICPDR and the BSC agreed to take measures to reduce the discharged loads of nutrients and hazardous substances to such levels necessary to permit the Black Sea ecosystems to recover to conditions similar to those observed in the 1960s. In line with this agreement, an annual exchange is conducted on observed river loads to the Black Sea (reported by the ICPDR based on measured data of discharges and concentrations of selected substances) and the status of the Black Sea ecosystem (reported by the BSC based on selected indicator parameters). In addition, the ICPDR regularly assesses the sources and pathways of nutrient emissions in the DRB and analyses the impacts of management and climate change scenarios on emissions and river loads to the Black Sea. The ICPDR and the BSC have also committed to adopt and enhance key water management strategies to ensure the implementation of appropriate practices and measures to reduce pollution and rehabilitate ecosystems.

Challenges

The ecological richness of the DRB and Black Sea is threatened by pressures exerted by human activities including urbanization, industrial development and agricultural practices, all of which are exacerbated by the impacts of climate change. The DRB is highly diverse showing a high variability of natural factors, socioeconomic conditions and cultural habits. Managing the Danube's resources in a sustainable manner requires collaborative efforts that embrace the basin's heterogeneity.

There are various challenges that require basin-wide coordination and cooperation between the ICPDR and BSC. The sustainable management of nutrients, hazardous substances and plastics in the basin is of particular importance for the Black Sea, being the ultimate recipient water body of the Danube and the world's most isolated sea. Other issues include the migration of fish species such as sturgeons from the Black Sea to the Danube, and severe droughts, which negatively affect the flow regime of the Danube, water-dependent economic sectors and aquatic ecosystems.

Results

In the late 1980s, the northwest shelf of the Black Sea was suffering from severe eutrophication due to high nutrient loads transported by the Danube. Since then, nutrient loads entering from the Danube have dropped significantly, and are currently approaching the ecological targets of the Black Sea. The situation in the shallow waters of the northwest Black Sea has improved considerably due to reduced nutrient inputs, resulting in a relatively stable trend of reduced eutrophication. The concerted actions of the Danube countries over the past 25–30 years to improve wastewater management and agricultural practices, in combination with a decline in the intensity of certain industries and agriculture in several countries, has led to a remarkable recovery.

Lessons learned

The inter-institutional collaboration extends beyond borders as the ICPDR plays a pivotal role in facilitating cooperation between the Danube countries and the Black Sea region. Established international legal mandates and well-structured cooperation among riparian countries is therefore a prerequisite for reaching environmental targets in transboundary basins.

Close transboundary cooperation can ensure coordinated technical work and harmonized policymaking, leading to agreed actions for joint implementation. Such processes should be supported by mutually verified and trusted data and assessment tools capable of fostering the development of national policies and measures. Active collaboration with appropriate international entities and related sectors, and the involvement of stakeholders and civil society, are key to proactively tackling novel water management challenges as they arise.

Source-to-sea management offers the possibility of implementing a holistic and integrated water management in the DRB, able to address pollution reduction, climate change adaptation and ecosystem preservation. The same approach should be applied to other river basins discharging into the Black Sea.

The current favourable situation should be maintained through the sustainable management of nutrients in the DRB and continued efforts to keep emissions low. Building resilience to climate change impacts on nutrient management is also key to address the negative consequences of extreme events on emissions. Implementation, maintenance and adaptation of measures should continue in the wastewater and agriculture sectors, among others, with an emphasis on enhanced wastewater treatment technologies, smart agri-environmental practices and nature-based solutions. The ICPDR supports riparian countries in these efforts by developing strategies and guidance documents, providing harmonized data and analytical tools, and fostering knowledge exchange and capacity building.

Source: Case Study provided by Birgit Vogel and Adam Kovacs, ICPDR, 2025.

3.2.2 Transboundary cooperation in coastal and marine areas

3.2.2.1 Regional Seas Conventions and Action Plans (RSCAP)

The Regional Seas Programme was established by UNEP in 1974, and has since become one of the effective international cooperation initiatives for the protection of the marine and coastal environment. The framework is built on legally binding conventions, often supplemented by protocols targeting specific issues like pollution control, conservation of marine ecosystems and management of marine resources. Currently, more than 146 countries participate in 18 RSCAPs. The Regional Seas Conventions under UNEP are a vital component of the global effort to protect and sustainably manage marine and coastal environments. UNEP's Regional Seas Programme consists of three types of RSCAP: UNEP Administrated, Non-UNEP Administered and independent (Table 3).

Table 3 Regional Seas Conventions and Action Plans

Administered by UNEP	• Convention for the Protection and Development of the Marine Environment of the Wider Caribbean Region (Cartagena Convention) • Framework Convention for the Protection of the Marine Environment of the Caspian Sea (Tehran Convention) • Action Plan for the Protection and Development of the Marine Environment and Coastal Areas of the East Asian Region (East Asian Seas Action Plan) • Nairobi Convention for the Protection, Management and Development of Coastal and Marine Environment of the Western Indian Ocean (WIO) region (Nairobi Convention) • Convention for the Protection of the Mediterranean Sea Against Pollution (Barcelona Convention) • Northwest Pacific Action Plan (NOWPAP) • Convention for Cooperation in the Protection, Management and Development of the Marine and Coastal Environment of the Atlantic Coast of the West and Central Africa Region (Abidjan Convention)
Independent but cooperating and coordinating with UNEP	• Convention on the Protection of the Black Sea Against Pollution (Bucharest Convention) • Kuwait Regional Convention for Cooperation on the Protection of the Marine Environment from Pollution (Kuwait Convention) • Convention for Cooperation in the Protection and Sustainable Development of the Marine and Coastal Environment of the North-East Pacific (Antigua Convention) • Convention for the Protection of the Natural Resources and Environment of the South Pacific Region (Noumea Convention) • Colombo Declaration on the South Asia Co-operative Environment Programme • Action Plan for the Protection of the Marine Environment and Coastal Areas of the Southeast Pacific • Regional Convention for the Conservation of the Red Sea and Gulf of Aden Environment (Jeddah Convention)
Independent	• 1991 Arctic Environmental Protection Strategy (Protection of the Arctic Marine Environment – PAME) • Convention on the Conservation of Antarctic Marine Living Resources (CCAMLR) • Convention on the Protection of the Marine Environment of the Baltic Sea Area (Helsinki Convention) • Convention for the Protection of the Marine Environment of the North-East Atlantic (OSPAR Convention)

It is important to note that each Regional Sea Convention focuses on a particular geographic area, with countries working together to implement coordinated actions and solutions for regional needs and environmental priorities stemming from their specific regions, *inter alia*, pollution, overfishing, habitat destruction, biodiversity loss and others. (Table 4). They are implemented by establishing regional or national policy tools such as Regional Action Plans and National Action Plans, and can also include transformative actions in their work programmes such as nature-based solutions and the blue economy.

© Lukas / Adobe Stock

Table 4 Examples of the relevance of Regional Seas Conventions and Action Plans to source-to-sea management

Convention	Source-to-sea relevance
Kuwait Convention	• 1990 Kuwait Protocol addresses marine pollution from land-based sources
Nairobi Convention	• 1996 Protocol on Land-Based Sources and Activities (LBSA Protocol)
East Asian Seas Action Plan	• 2019 Regional Action Plan on Marine Litter under the Coordinating Body on the Seas of East Asia: Consolidate, coordinate and facilitate cooperation, and implement the necessary environmental policies, strategies and measures for sustainable, integrated management of marine litter including from land-based activities and across the full lifecycle
Tehran Convention	• 1995 Protocol Concerning the Protection of the Caspian Sea against Pollution from Land-based Sources and Activities (Moscow Protocol) • 2014 Protocol on Biodiversity Conservation • 2018 Protocol on Environmental Impact Assessment in a Transboundary Context
OSPAR Convention	• 1998 OSPAR Decision 98/4 on Emission and Discharge Limit Values for the Manufacture of Vinyl Chloride Monomer (VCM) including the Manufacture of 1,2-dichloroethane (EDC) from point and diffuse sources
Helsinki Convention	• Convention covers the whole catchment and its inland waters and obliges Contracting Parties to take joint measures also in transboundary watercourses • Working Group on Source to Sea Management of Nutrients and Hazardous Substances and Sustainable Agricultural Practices oversees related recommendations, implementation of actions and carries out joint activities with transboundary River Commissions
Barcelona Convention	• 1980 Protocol for the Protection of the Mediterranean Sea against Pollution from Land Based Sources and Activities (LBS Protocol) (amended in 1996) • Regional Plan on Marine Litter (Decision IG 25/9) is designed to prevent and reduce marine litter pollution across the Mediterranean Sea • Regional Plan on Urban Wastewater Treatment (Decision IG 25/11) aims to safeguard coastal and marine environments, as well as human health, from the harmful effects of both direct and indirect discharges of untreated wastewater • Regional Plan on Agriculture (Decision IG 26/6) focuses on reducing and preventing pollution arising from the use of fertilizers, pesticides and agricultural waste.

Catchment and inland waters-related provisions can be included directly in Regional Seas Conventions. For instance, Article 6 Principles and obligations concerning pollution from land-based sources in the Helsinki Convention calls for Contracting Parties to prevent and eliminate pollution of the Baltic Sea area from land-based sources by using best environmental practices for all sources and best available technologies for point sources. Article 6 also calls on Contracting Parties to implement procedures and measures concerning emissions and inputs to water and air, environmental quality, and products containing harmful substances and materials, as well as prevention of pollution from agriculture. The HELCOM Working Group on Source to Sea Management of Nutrients and Hazardous Substances and Sustainable Agricultural Practices (WG Source to Sea) works towards reducing pressures on the marine environment, monitors progress towards reaching related goals, implements relevant mitigation measures, and monitors the state of the sea in relation to these aspects (Case Study 11).

11 Case Study

Cooperation on target setting and pollution prevention in the Baltic Sea Basin

Context

The Baltic Marine Environment Protection Commission (HELCOM) Working Group on Source to Sea Management of Nutrients and Hazardous Substances and Sustainable Agricultural Practices (WG Source to Sea) works towards reducing pressures on the marine environment, consisting primarily of inputs of nutrients and hazardous substances to the Baltic Sea; monitoring progress towards the goal of a Baltic Sea unaffected by eutrophication and hazardous substances, as enshrined in the 2021 Baltic Sea Action Plan (BSAP); and implementing relevant mitigation measures.

Activities

Mitigation measures that embody the source-to-sea approach include the HELCOM Recommendations on wastewater management, sewage sludge management, stormwater management, mitigating ammonia emissions from agriculture and the prevention of pollution from land-based sources. Practical efforts to address catchment pollution, including transboundary pollution, include the compilation of Pollution Load Data (PLC). These data have been an integral part of the HELCOM assessment system since 1987. This system comprises annual and periodic assessments of nutrient inputs and selected hazardous substances, based on national monitoring and modelling in accordance with joint regional guidelines and strategies.

Challenges

The main challenge for HELCOM, as an intergovernmental organization, is reaching subregional and local stakeholders – despite having over 60 regional Observer organizations representing local governments, non-governmental organizations and interest organizations for different sectors. Observers are by definition regional actors on the Baltic Sea level, thus actual local or single stakeholders are only represented in organizations at the regional level.

Political support and commitment on the part of Contracting Parties exists for agreed actions, but additional funding and project formation is a persistent requirement, in order to implement joint and national actions outlined in the 2021 BSAP – for example, to reduce nutrient, hazardous substance or litter inputs.

Results

As a prime example, HELCOM's regional work on reducing catchment and airborne inputs of nutrients which cause eutrophication has resulted in significant decreases in nitrogen, and especially phosphorus, entering the sea over decades. With the support of the HELCOM Recommendations, programmes including the regional hot spots approach, and projects, the phosphorus load to the sea has been cut by c. 50,000 tonnes from its peak of over 71,000 tonnes at the beginning of 1980s, to less than 23,000 tonnes. Recent reductions from the reference period of 1997–2003 to the implementation of the regional nutrient reduction scheme have been 28 per cent for phosphorus and 12 per cent for nitrogen.

The keys to the success of this component of HELCOM's policy work, aside from the joint BSAP (2007 and revised 2021), are regional nutrient reduction targets (2007), the regional action plan on marine litter (2015), and the political will and commitment to implement these intergovernmental objectives.

Lessons learned

Jointly agreed monitoring methods and practices, joint reporting and data management are needed as well as a sound and robust scientific basis and modelling. Methods and tools for regional assessments, with clear and transparent joint tracking of progress of reduction targets and other management objectives, are an important part of the work and essential to showcasing the results of joint action.

HELCOM is the key management framework for addressing pressures and activities, and is also applied to hazardous substances, though related measures, monitoring and reporting are still under development.

A new regional strategic approach on managing hazardous substances was developed over two years, and once the tools were ready, an initial test run was conducted. The involvement of externally funded regional projects and experts in the HELCOM community enabled the development of a new mechanism to produce and manage HELCOM's list of priority substances, as well as their sources and pathways, and helped identify gaps with a view to preparing a new HELCOM Recommendation for implementation in the coming years.

To reach the nutrient reduction targets, effective implementation of the regional nutrient recycling strategy, as well as increased engagement with upstream countries beyond HELCOM Contracting Parties, will be needed. Source-to-sea management of the Baltic Sea basin, and the work of the WG Source to Sea, is therefore in a state of continuous development and improvement.

Source: Case Study provided by Lotta Ruokanen, HELCOM, 2025.

3.2.2.2 Global Programme of Action (GPA)

UNEP's GPA aims at preventing the degradation of the marine environment from land-based activities. The GPA specifically addresses land systems, freshwater systems, estuaries and coastal areas, and the connectivity between them. It helps States act alone or together, according to their policies, priorities and resources, to prevent, reduce, control and/or eliminate marine pollution and support ecosystem recovery. Its scope includes sources of sewage, POPs, radioactive substances, heavy metals, oils (hydrocarbons), nutrients, sediment mobilization, litter, and the physical alteration and destruction of habitats.

The work of the GPA is carried out primarily through three sister partnerships, namely: the Global Partnership on Plastic Pollution and Marine Litter (GPML), the Global Wastewater Initiative (GWWI) and the Global Partnership on Nutrient Management (GPNM).

The GPA has partnered with RSCAPs on Protocols on Land-Based Activities (LBA), monitoring programmes, LBA assessments, and regional and national plans. The programme, via RSCAPs, regulates pollution flows (nutrients, plastics including marine litter and chemicals) from land-based sources into the marine environment, as well as material and sediment flows that impact estuaries. Strategic partners and stakeholders are defined as national governments, regional organizations, the scientific community, the private sector, civil society, and other United Nations entities and international organizations. The GPA engages with stakeholders and strategic partners through national governments. The GPA also facilitates cooperation across regions through MOUs, joint actions and regional implementation strategies, working in alignment with other regional conventions.

3.3 Opportunities for incorporating source-to-sea management in transboundary cooperation

Diverse legal and institutional arrangements can support the development of cooperation between transboundary water agreements and RSCAPs and their related operational organizations. Incorporating source-to-sea management into existing cooperation endeavours can enlarge the suite of benefits that can be exchanged in a transboundary setting. It can also reduce risks by considering a larger universe of costs and benefits, and avoid unintended consequences by highlighting impacts that can affect downstream, as well as upstream, ecosystems and communities. While this is a growing trend, the third report on the progress on transboundary water cooperation under the Water Convention 2020–2023 notes that when asked whether a joint body had ever invited a non-riparian coastal State to cooperate, an overwhelming majority of responses indicated this was not the case (UNECE, 2024a).

Coordination between joint bodies can lead to well-balanced measures that avoid negative social, environmental and/or economic impacts on downstream coastal and marine areas, but also in upstream parts of river systems. It can consolidate resources and reduce costs by focusing on systemwide benefits, thereby increasing the effectiveness and efficiency of actions taken (Case Study 12).

12 Case Study

Cooperation on joint activities for improvements from the Lesotho Highlands to the Atlantic Ocean

Context

In Southern Africa, effective source-to-sea governance of transboundary water systems and coastal ecosystems is essential to ensure the sustainable development and preservation of vital, natural resources for people, ecosystems and countries. A key area of focus is the integrated management of the Orange River Basin and the Benguela Current ecosystem. The collaboration between the Orange-Senqu River Commission (ORASECOM) and the Benguela Current Commission (BCC) plays a critical role in coordinating efforts to protect and sustainably manage these interconnected ecosystems.

ORASECOM was established in 2000 to promote integrated water resources management (IWRM) in the Orange River Basin. ORASECOM works to implement strategies that balance economic development, environmental sustainability and social equity. By coordinating the efforts of its member countries, ORASECOM helps prevent disputes over water rights and fosters collaborative projects to improve water management across the basin. The BCC was established in 2007 to promote the sustainable management of the Benguela Current ecosystem and the resources it supports. The BCC works to protect the health of the Benguela Current through collaborative monitoring, research and conservation initiatives. By bringing together governments, scientists and other stakeholders, the BCC addresses challenges such as overfishing, habitat degradation and the impacts of climate change on the marine environment.

Lessons learned

The integrated management of the Orange River Basin and the Benguela Current is an example of source-to-sea management, which recognizes the interconnectedness of freshwater and marine ecosystems. The collaboration between ORASECOM and the BCC is essential to managing the full course of the Orange River – from its source in the Lesotho Highlands to its mouth in the Atlantic Ocean. This cooperation ensures that water quality and quantity in the river basin are maintained in a way that minimizes negative impacts on the marine environment. Joint efforts include monitoring water quality, implementing sustainable agricultural practices and mitigating pollution, particularly from mining and industrial activities along the river.

Source: Case Study provided by Susanne Schmeier, IHE Delft and Scientific and Technical Advisory Panel (STAP) of the Global Environment Facility (GEF), 2025.

Effective transboundary cooperation for source-to-sea management can be facilitated through various modalities of cooperation between transboundary and regional seas conventions and joint bodies. Legal arrangements can include Memorandums of Understanding (MOUs), Letters of Agreement and shared Protocols, while cooperation can be facilitated through Joint Task Forces (JTFs), Expert Groups and Working Groups (WGs).

The legal, institutional and practical implementation linkages between treaties and agreements, RBOs and RSCAPs, can provide opportunities to progress towards source-to-sea management. For example, impacts on the Black Sea from changes in key source-to-sea flows, such as water, pollutant and sediment flows in the Danube River Basin, has led to cooperation between the ICPDR and ICPBS since 2000. In 2001, they signed an MoU on common strategic goals (ICPDR, 2001) outlining the long-term objective of taking measures to reduce the loads of nutrients and hazardous substances to such levels necessary for Black Sea ecosystems to recover.

The MoU also provides the institutional implementation framework supporting scientific and technical cooperation between these two bodies for action toward these objectives. Data from the Transnational Monitoring Network for the period 2008–2022 indicate that fluxes of dissolved inorganic nitrogen and total phosphorus transported by the Danube River to the Black Sea have been reduced by 40 per cent from levels in the 1980s. In the Danube Declaration (ICPDR, 2022) adopted at the ICPDR Ministerial Meeting in 2022, the Ministers reiterated the importance of combined efforts to preserve the Black Sea ecosystem and cooperation with ICPBS.

At the 23rd Meeting of the Contracting Parties to the Barcelona Convention (COP 23) in Portorož, Slovenia, held in December 2023, an agreement was concluded to strengthen as well as intensify cooperation between the Convention for the Protection of Danube River, the Black Sea Convention and the Barcelona Convention, with a view to enhancing implementation of a joint source-to-sea approach designed to reduce pressures and impacts from land-based sources on the coastal and marine environment of the ecosystem shared between the Danube Basin, the Black Sea and the Mediterranean Sea.

As of spring 2025, discussions are ongoing to develop an MoU between the Barcelona Convention, the ICPDR and the ICPBS to intensify transregional cooperation and strengthen joint source-to-sea management by harmonizing future activities and promoting synergies to promote sustainable blue and green economies within the Danube, Black Sea and Mediterranean basins.

© chokchaipoo / Adobe Stock

As noted above, MOUs can be used to formalize cooperation on source-to-sea management. For example, in the Drin River Basin, the establishment of Drin Coordinated Action (Drin CORDA) by the riparian countries in 2011 (Case Study 13) created a framework for the implementation of a Memorandum of Understanding for the Management of the Extended Transboundary Drin Basin (Drin MoU). The Preamble of the MOU affirms that "*the Drin River is the connecting agent of an extended shared watershed, including a number of shared water bodies and an adjacent sea, the Adriatic, linking these into a hydrologic system that supports a variety of ecosystems within the Drin Basin*". Article 4 of the MoU then calls for priority actions at national, bilateral and/or multilateral levels including "No Regret" measures to integrate the consideration of coastal ecosystems, transitional waters and shallow marine environment.

13 Case Study

Developing upstream-downstream and cross-country cooperation for sustainable source-to-sea management of the Drin River Basin

Context

The Drin River originates in Lake Prespa in northern Greece, then flows north into the Adriatic Sea where it joins the Buna/Bojana river flowing out of Lake Skadar/Shkodër. Its basin is shared by Albania, Greece, Kosovo (United Nations administered region, Security Council resolution 1244 (1999)), Montenegro and North Macedonia. The Drin River Basin constitutes a source-to-sea system encompassing the river basin itself (including its tributaries and lakes), connected aquifers, downstream recipients (Buna/Bojana Delta and estuaries), and the coastal and adjacent marine environments.

Activities in the upper reaches of the Drin basin directly impact downstream areas, including Lake Shkodra/Skadar, the Buna/Bojana River and the Adriatic Sea. Sediment transport, nutrient flows and pollutant discharge are all continuous processes that affect the entire system. Source-to-sea management is vital for ensuring that the transboundary management of the Drin basin is comprehensive, effective and sustainable, safeguarding the health of both freshwater and marine ecosystems. Source-to-sea management enables spatial integration in the development of policy and management instruments, enhancing cooperation among the Drin riparian countries.

Key flows

The Drin Transboundary Diagnostics Analysis (TDA) was completed in 2020, leading to the endorsement of the Drin Strategic Action Programme (SAP) by ministers and high-level representatives of the Drin riparian countries on 24 April 2020.

The management of the Drin Basin involves considering several interconnected flows:

- **Water.** Managing water quantity and flow regimes is crucial to address flood risks in the downstream and coastal parts of the basin, as well as for agriculture, hydropower and the health of freshwater-dependent and marine ecosystems. Observed shifts in water flow patterns in the basin are due to climate variability and changes, hydropower generation and seasonal over-abstraction of water resources.
- **Sediment.** Natural sediment transport is vital for maintaining coastal ecosystems. Disturbances in the transport regime are caused mainly by hydropower dams, gravel extraction and deforestation. Key impacts include decreased sediment loads and increased erosion downstream of hydropower plants, coastal changes from changing sediment processes and interaction with the sea, and loss of habitats from changes in sediment deposition.
- **Pollution.** Surface water, groundwater and marine waters are all affected by pollution in the Drin basin coming from various sources including domestic wastewater, agricultural runoff, industrial discharges, mining activities and solid waste.
- **Biota.** Biodiversity degradation, including in coastal habitats, is due mainly to pollution, changes in the hydrological regime and illegal activities (e.g. illegal fishing).

Mechanism for coordination

Drin Coordinated Action (Drin CORDA) was established in 2011 following a multi-stakeholder policy dialogue initiated in 2009, and preparatory technical and political engagement work undertaken since 2006. This framework, established by the Drin riparian countries for implementation of the Drin MoU, was signed by the ministers responsible for management of water resources and/or the environment, and high-level representatives of the riparian States in Tirana, on 25 November 2011.

The governments of the Drin riparian countries have established the Drin Core Group (DCG) and its Expert Working Groups to coordinate implementation of the Drin MoU. The DCG operates as a de facto joint commission for management of the Drin Basin.

Key benefits realized under the Drin Coordinated Action Process include:

- enhanced consensus among countries on key transboundary concerns and drivers of change from source to sea;
- agreement on a programme of priority actions to address transboundary concerns and achieve the shared vision;
- strengthening of technical and institutional capacities;
- piloting of novel management approaches with potential for replication and upscaling, including wastewater treatment (through nature-based solutions), design of flood insurance schemes as flood risk transfer mechanisms, and biomass collection and production of fuel briquettes;
- increased sense of ownership of the process and results among stakeholders;
- increased understanding of the need for sectoral and spatial integration for policymaking and management of natural resources.

Challenges

The main challenges or barriers to source-to-sea management are sectoral approaches and policies regarding the management of natural resources. Upstream-downstream considerations for management of the basin, the coastal zone and the marine area are not taken into consideration in the development of policy instruments and management tools, both at the national and transboundary levels. Although financing is available through the donor community and the European Commission (via the EU accession process), there is little coordination among efforts and projects. Much of the financial and technical support provided is driven by donor policies, exacerbating the adverse effects of non-integrated natural resources management. In addition, technical and scientific capacities for informed decision-making are restricted.

Lessons learned

The Drin River, lakes and supporting ecosystems extend and connect multiple countries. Implementing source-to-sea management along with other approaches such as integrated water resources management (IWRM), Water-Energy-Food-Ecosystem (WEFE) Nexus and ecosystem-based approaches enables the development of policy instruments and management tools at the national and transboundary levels. These are scaled to address upstream/downstream effects across ecosystems and countries in ways that take into consideration sectoral considerations. A clear understanding of these approaches and the benefits that stem from their application is crucial.

Next steps

The steps foreseen for the next five years are designed to improve and scale up source-to-sea management efforts. Key benefits expected through planned activities include:

- the development of technical and policy tools to fill gaps in the understanding of the functioning of Drin Basin ecosystems across the source-to-sea continuum through the development of an updated TDA, the design of multi-purpose transboundary monitoring programmes, assessments of the status of freshwater-dependent ecosystems as well as of minimum ecosystem flows, the establishment of an improved information management system, and the training of staff of institutions and other stakeholders;
- the establishment of a transboundary institutional arrangement and the development of critical transboundary policy instruments through the development of a legal instrument consolidating official cooperation among the Drin riparian countries, the establishment and operation of a joint coordination mechanism (Drin Commission), and the development of a River Basin Management Plan for the Drin system;
- the implementation of solutions to address main causes of transboundary concern and upstream/downstream effects including small-scale wastewater treatment plants in small rural settlements, sustainable agricultural practices for nutrient and toxic pollution reduction, erosion control in priority areas, support to SMEs to facilitate transition of the local economy towards sustainable natural resources management practices and green growth, and sustainable transboundary fisheries management.

Source: Case Study provided by Tassos Krommydas, Global Water Partnership-MED, 2025.

Cooperation on source-to-sea management can also be incorporated into transboundary arrangements developed after the original transboundary agreement has already been in force for years or even decades. For example, riparian countries can agree upon a Protocol to an agreement that further defines the areas of cooperation under the initial agreement. A Protocol focusing on one or more aspects of source-to-sea management could strengthen cooperation for the benefit of the entire source-to-sea system (UNECE, 2013b). For example, Germany and the Netherlands have agreed to a supplementary environment Protocol under the Ems-Dollard Treaty, which deals with cooperation in the water and nature conservation sectors, in particular in the areas of water quality and water ecology.

RSCAPs frequently have specific areas which are regulated by protocols that constitute an integral part of the Convention. Environmental objectives can be established in a protocol or an annex to regional seas conventions – for example, the "Protocol for the Protection of the Mediterranean Sea against Pollution from Land-Based Sources and Activities (LBS Protocol)" in the case of the Barcelona Convention (1976). The protocol requires countries to develop national and regional action plans that contain measures and timetables on minimizing pollution from various land-based activities.

Contracting Parties to this Protocol are obligated to "*take all appropriate measures to prevent, abate, combat and eliminate to the fullest possible extent pollution of the Mediterranean Sea Area caused by discharges from rivers, coastal establishments or outfalls, or emanating from any other land-based sources and activities within their territories, giving priority to the phasing out of inputs of substances that are toxic, persistent and liable to bioaccumulate.*"

To achieve this, as articulated in Article 5, Parties individually or jointly elaborate and implement national and regional action plans and programmes, such as the Regional Plan on Marine Litter (Decision IG 25/9) aimed at preventing and reducing marine litter pollution across the Mediterranean Sea, or the Regional Plan on Urban Wastewater Treatment (Decision IG 25/11) designed to safeguard the coastal and marine environments, as well as human health, from the harmful effects of both direct and indirect discharges wastewater. In addition, Article 11 of the protocol specifically requires any Party that is the riparian of a transboundary river flowing into the Mediterranean Sea, to cooperate with upstream riparian States on pollution reduction, even when the upstream riparian States are not a party to the protocol.

In line with the aims of the LBS Protocol (Barcelona Convention), States can undertake activities within their boundaries to reduce the impacts of land-based activities on coastal and marine areas. For example, the MedProgramme, facilitated by GEF funding, was designed to support implementation of the Barcelona Convention within the framework of its Medium-Term Strategy, as well as facilitate the adoption and implementation of relevant Protocols of the Convention. MedProgramme activities are embedded in a solid legal and institutional cooperation framework to ensure policy coherence of actions implemented in different segments of the source-to-sea continuum and across different sectors and countries.

Joint projects, in the absence of an agreement or complementing an agreement which does not include source-to-sea aspects, can strengthen cooperation between States while creating benefits for the transboundary system. Implementation of projects and activities at the national and subnational levels can support and reinforce source-to-sea management (see Case Study 2, Case Study 3, Case Study 7, Case Study 8, Case Study 14, Case Study 18, Case Study 19, Case Study 21Case Study 22, Case Study 24, Case Study 25 and Case Study 26). These activities contribute social, environmental and economic benefits to the source-to-system and can provide examples of interventions that can be scaled up in the transboundary context for greater impact.

Working groups can operationalize the aims of legal agreements. For example, HELCOM has eight main working groups which handle specific topics related to the Baltic Sea's environment, catchment or maritime activities. The groups are major contributors to the HELCOM work programme, gathering science and technical expertise and translating their findings into policies, strategies or recommendations best suited to respond to a particular issue affecting the Baltic Sea. HELCOM WG Source to Sea works to reduce pressures on the marine environment – primarily inputs of nutrients and hazardous substances to the Baltic Sea – to monitor progress towards achieving the goal of a Baltic Sea unaffected by eutrophication and hazardous substances, and to implement relevant mitigation measures.

3.3.1 Cooperation on joint action plans

Source-to-sea considerations can be incorporated into joint action plans between riparian and coastal States or between RBOs and RSCAPs. The action plans can prioritize actions that States take to comply with multiple legal frameworks related to the land, freshwater, coastal and marine environments, and the flows that link them. These action plans can be supported with concrete financial instruments that incentivize source-to-sea action. Cooperation can enhance coordination between national legal frameworks and strategic documents, and between various management plans along the source-to-sea continuum.

The revised 2021 Baltic Sea Action Plan (BSAP) lays out strategic regional policy measures for the protection of the marine environment of the Baltic Sea. It contains 199 actions including, for example, concrete environmental targets in the form of maximum allowable inputs (MAIs) for sub-basins and some major rivers, national nutrient input ceilings (NICs) and 36 actions to address eutrophication.

In the case of the Western Indian Ocean, the 10 countries developed the Strategic Action Programme for the Protection of the Western Indian Ocean from Land-Based Sources and Activities (WIOSAP). This programme followed completion of the Transboundary Diagnostic Analysis, a process used by the GEF to build a shared knowledge base of the biophysical, social and governance context. Through implementing a variety of projects, WIOSAP aims to reduce land-based stressors of coastal and marine ecosystems by protecting critical habitats, improving water quality, managing river flows, and promoting regional collaboration and governance frameworks that prevent ecosystem degradation (Case Study 14).

14 Case Study
Community-based ecological coastal rehabilitation using an ecosystem approach in the Seychelles

Context

As part of the WIOSAP programme under the Nairobi Convention, the Seychelles Wetland Rehabilitation project focused on restoring degraded wetlands and foothills in Pasquière, Praslin Island, Seychelles. The source-to-sea system being managed encompasses the Pasquière, Davidson and Casimir rivers, freshwater marshes, mangrove forests and the marine park lagoon. Natural solutions, known as ecosystem-based approaches, were used by local non-governmental organizations and communities to rehabilitate the degraded coastal ecosystems.

Project partners included the Waste Enforcement and Permits Division, the Environment Department, the Ministry of Environment, Energy and Climate Change, Gaea Conservation Network Seychelles, the Chinese Academy of Sciences, the University of Seychelles Centre for Environmental Education, the Praslin Watershed Committee and BS Excavation. WIOSAP funds were leveraged to receive funds from other donors (GOSUNDP-GEF Ridge to Reef, SADC GCCA+ and the Global Fund for Coral Reef/SeyCCAT project).

Activities

The project rehabilitated 3 hectares of wetland and associated habitats, restored 2 hectares of a degraded hill by planting 6,000 tree seedlings and integrated an innovative palm-leaf anti-erosion barrier. These hydrological restoration efforts have enhanced water flow, benefiting both wetland ecosystems and adjacent marine habitats, including seagrass beds and coral reefs.

Lessons learned

The project successfully demonstrated that ecosystem-based adaptation approaches for ecosystem restoration could work, and that this demonstration project should be upscaled and replicated with more innovative ideas. Training and capacity-building efforts are important for long-term sustainability and local ownership of wetland management. Collaboration with government agencies and universities ensures that scientific data informs management decisions and policy recommendations, enabling more effective and scalable rehabilitation efforts.

Source: Case Study provided by Agnes Mukami Muriuki, UNEP, 2025.

Inside tree nursery ran by TRASS at Praslin Island. Trees from nursery are used to rehabilitated degraded wetlands © Melisa Wandia / Nairobi Convention

3.3.2 Cooperation on setting targets/pollution prevention

Setting pollution control targets and establishing common criteria for water quality standards and sediment control are essential for successful source-to-sea management. Article 3 of the Water Convention provides a legal basis for setting pollution control standards and water quality criteria in transboundary waters. The Parties are required to establish effective measures to prevent and reduce pollution from point sources (e.g. industries, urban wastewater) and diffuse sources (e.g. agriculture), ensuring that these measures are implemented upstream to protect downstream aquatic and marine ecosystems. Article 9 (2) (e) goes further by asking joint bodies to elaborate joint water quality objectives and criteria and, where necessary, to propose relevant measures for maintaining and, where necessary, improving existing water quality. Moreover, Article 12 requires riparian Parties to undertake research and development activities in support of achieving water quality objectives and criteria.

The OSPAR Convention protects the marine environment of the North-East Atlantic, focusing on pollution control, habitat conservation and the sustainable management of marine resources. OSPAR covers coastal areas, the sea shelf and the open sea. It regulates water flow, pollution and sediment in the North-East Atlantic, with a strong emphasis on the management of hazardous substances, oil spills and marine litter. Stakeholders are defined as member countries, environmental organizations and industrial sectors (e.g. shipping, fisheries). OSPAR engages stakeholders directly through national bodies and indirectly through regulatory enforcement by national agencies.

OSPAR encourages cooperation between member countries through joint protocols, shared monitoring programmes and MOUs aimed at protecting the North-East Atlantic. Article 3 on Land Based Sources and Activities specifically targets pollutants, including heavy metals, POPs and nutrients from land-based activities. Parties are obliged to take preventive measures to reduce or eliminate hazardous substances entering the marine environment, with a clear emphasis on pollution prevention in upstream sources, especially agriculture and urban wastewater discharges.

In the Cartagena Convention, Article 7 addresses pollution from land-based sources, stating that *"Contracting Parties shall take all appropriate measures to prevent, reduce and control pollution of the Convention area caused by coastal disposal or by discharges emanating from rivers, estuaries, coastal establishments, outfall structures or any other sources on their territories"*.

Article 8 of the Barcelona Convention focuses on preventing land-based sources of pollution, particularly those related to discharges directly into the sea or through coastal outfalls or indirectly from the rivers, canals or other watercourses including underground water courses or run-off, from affecting the Mediterranean Sea. It thus emphasizes the importance of upstream pollution prevention. Accordingly, Article 6 of the LBS Protocol requires authorizations and regulating systems for any release to the water systems and air. Article 7 of the LBS Protocol calls for, among others, the establishment of standards and criteria for effluent discharges to the water and marine segments.

The GEF-8 Clean and Healthy Ocean Integrated Program (CHO-IP) aims to tackle marine hypoxic zones by curbing coastal pollution from agriculture, industrial and municipal sources through policy and regulatory measures and infrastructure investments combined with nature-based solutions. The programme fosters an enabling environment to prevent and alleviate the damaging impacts of eutrophication and hypoxia in coastal and marine environments, thereby advancing the associated ecological, social and economic well-being of countries and Large Marine Ecosystems (LMEs). By linking upstream nutrient sources to downstream ocean health, CHO-IP helps countries deliver on blue economy goals and achieve SDG 6 (water), SDG 14 (life below water) and SDG 13 (climate action) (Case Study 15).

15 Case Study

Curbing land-based pollution to address marine hypoxia: the GEF-8 Clean and Healthy Ocean Integrated Program (CHO-IP)

Challenge

Marine hypoxia – oxygen depletion in coastal and open waters – is a worsening global crisis driven by land-based nutrient pollution and climate change. Excess nitrogen and phosphorus from agriculture, industry and municipal sources flow through river basins into estuaries and coastal zones, triggering eutrophication, harmful algal blooms and oxygen loss. These processes disrupt biogeochemical cycles and create "dead zones" (oxygen levels under 2 mg/L) that severely degrade biodiversity, fisheries and ecosystem services. As these hypoxic zones expand, so too does the risk to food security, livelihoods and sustainable development.

Key drivers and barriers

The agriculture and livestock sectors are the largest contributors to nutrient pollution globally, followed by rising contributions from urban and industrial waste. A lack of awareness, ineffective policy frameworks and limited access to finance and innovation hamper efforts to curb these sources. Three core barriers have been identified:

- insufficient knowledge, monitoring and awareness of hypoxia causes and consequences;
- weak policy coherence, inadequate regulation and poor investment in pollution reduction;
- limited stakeholder access to best practices, nature-based solutions (NBS) and enabling financial mechanisms.

Objectives and strategy

CHO-IP aims to reduce nutrient pollution across the entire source-to-sea continuum by:

- addressing pollution at its source by supporting upstream interventions in agriculture, urban wastewater and industry to curb land-based nutrient inputs;
- enhancing ridge-to-reef connections by linking riverine nutrient transport to coastal hypoxia outcomes through integrated governance, planning and investment frameworks;
- supporting gender-responsive policy, strategy and regulatory frameworks with science-based information, targeted guidance, and analysis and decision-support tools;
- scaling nature-based solutions and best management practices;
- restoring the oxygen balance in degraded coastal zones in participating countries;
- engaging stakeholders across freshwater and marine systems to deliver coordinated, systemic change.

CHO-IP embeds a source-to-sea lens by delivering a comprehensive response to nutrient-driven marine hypoxia – restoring coastal ecosystems, improving water quality throughout watersheds, and enhancing resilience from catchment to coast to open ocean – through four integrated components:

- Awareness and knowledge (Component 1)
  - ✓ *builds a shared understanding across sectors and scales;*
  - ✓ *delivers tools for cooperative monitoring, early warning and science-policy engagement;*
  - ✓ *strengthens data systems connecting land and sea pressures.*
- Policy and investment (Component 2)
  - ✓ *aligns upstream and downstream policy frameworks;*
  - ✓ *creates financial mechanisms to support integrated nutrient reduction goals;*
  - ✓ *advances incentives for land-based pollution control and marine restoration.*
- Best practice upscaling (Component 3)
  - ✓ *demonstrates and scales nature-based solutions (e.g. buffer zones, wetland restoration) and improved waste practices;*
  - ✓ *focuses on reducing nutrient loading from farms, cities and industries.*
- Monitoring and coordination (Component 4)
  - ✓ *tracks impacts from ridge to reef;*
  - ✓ *ensures adaptive, evidence-based programme management and reporting;*
  - ✓ *supports stakeholder engagement (public and private sectors) across freshwater and marine domains;*
  - ✓ *mainstreams gender equality and inclusivity across the programme;*
  - ✓ *fosters programmatic coordination among child projects.*

Programmatic approach and partnerships:

CHO-IP is supported by the GEF; co-led by FAO, UNDP, Conservation International, the European Bank for Reconstruction and Development (EBRD), UNEP, the World Bank, the Inter-American Development Bank (IADB), the Asian Development Bank (ADB) and the Development Bank of Latin America (CAF); and co-executed by a large partnership of international organizations. The programme encompasses 15 Child Projects comprising 1 Global Coordination Project and 14 Country Projects: Grenada, Jordan, Madagascar, Maldives, Mexico, Moldova, Panama, Peru, Sri Lanka, St. Kitts and Nevis, Thailand, Trinidad and Tobago, Viet Nam and Venezuela. These represent 10 Large Marine Ecosystems: the Bay of Bengal (+ Gulf of Thailand + Arabian Sea), the South China Sea, the Caribbean (+ Gulf of Mexico), the Red Sea, the Black Sea, the Humboldt Current, the Agulhas Current and the Mediterranean Sea).

The Global Coordination Project (GCP) enhances the effectiveness of CHO-IP in addressing nutrient pollution and marine hypoxia through science-based knowledge, gender-responsive policies and regulatory measures, infrastructure investments, and inclusive awareness-raising initiatives globally, regionally and in participating countries, thereby delivering increased global environmental benefits. The GCP represents CHO-IP and facilitates programmatic coordination and reporting.

The total GEF grant amounts to $112 million, and co-financing is valued at $748 million. CHO-IP will start execution of its programme objectives in 2025 for the duration of six years.

Source: Case Study provided by Lucilla Minelli, FAO, 2025.

3.3.3 Cooperation on science and monitoring programmes

Monitoring plays a critical role in source-to-sea management, especially for understanding the social, environmental and economic linkages between segments, sources of alteration to key flows and their impacts, and for assessing the cumulative impacts of human activities across the source-to-sea system. Countries must cooperate in data-sharing and joint monitoring programmes to compile a comprehensive picture of the source-to-sea system (Case Study 1).

Article 11 of the Water Convention on Monitoring and Information Exchange mandates countries to establish systems for the monitoring of transboundary waters, exchange of data and joint assessments of water quality. For source-to-sea management, it is essential that monitoring focuses on the downstream impacts of upstream activities and ensures the protection of both freshwater and marine ecosystems. Value can also be gained by monitoring how impacts and benefits travel upstream, from sea to source.

Article 4 on Specific Provisions for Joint Management of Watercourses of the SADC Protocol on Shared Watercourses stresses the importance of joint monitoring and information sharing, particularly for managing water quality and sediment flow in shared watercourses. It also underscores the need for monitoring programmes that assess how upstream activities impact downstream aquatic and marine environments.

Article 6 of the OSPAR Convention on Cooperation on Monitoring and Assessment requires Parties to establish joint monitoring and assessment programmes to measure the health of the marine environment, including the impact of nutrient loading and pollution from freshwater sources. The monitoring programmes focus on the downstream marine environment, ensuring that pollution from land-based sources is tracked and mitigated.

Article 12 of the Barcelona Convention on Monitoring and Assessment of Pollution calls for the development of regional monitoring programmes to assess the levels of pollutants in the Mediterranean Sea, including those that are carried through freshwater systems. Downstream coastal and marine monitoring focuses on the impact of water quality and sediment flux on marine habitats and biodiversity. For example, the Integrated Monitoring and Assessment Programme (IMAP), conducted within the framework of the Barcelona Convention, enhances the collective knowledge of the marine and coastal environment and its interactions with human activities in the context of sustainable development. The programme is implemented through agreed Common Indicators accompanied by agreed guidance documents including sampling protocols and assessment criteria, developed by joint expert groups and approved by the Contracting Parties. Article 13 on scientific and technological cooperation encourages, to the extent possible, Parties to cooperate directly, or when appropriate through competent regional or other international organizations, in the fields of science and technology and to exchange data as well as other scientific information.

Article 24 of the Helsinki Convention, which covers scientific and technological cooperation, calls for Contracting Parties to undertake directly, or when appropriate through competent regional or other international organizations, to cooperate in the fields of science, technology and other research, and to exchange data and other scientific information, as well as to promote studies and to undertake, support or contribute to programmes aimed at developing methods assessing the nature and extent of pollution, pathways, exposures, risks and remedies in the Baltic Sea Area.

On the operational side, there are 12 agreed HELCOM monitoring programmes covering sources and inputs of human pressures and various variables reflecting the state of the environment. HELCOM monitoring programmes are compiled in the HELCOM Monitoring Manual. The HELCOM monitoring programmes are the source of data for indicator-based assessments of the state of and pressures on the marine environment, as well as the analysis of long-term trends. The first HELCOM period assessment was published in 1986. Today HELCOM regularly carries out thematic and holistic assessments to evaluate progress towards the goals and objectives set for the Baltic Sea environment. The HELCOM assessment products are the basis for sound decisions to restore the Baltic Sea ecosystem, to support implementation of HELCOM objectives and actions, and to identify emerging environmental problems. Coordinated monitoring programmes at the regional Baltic Sea level are central for these activities.

The eight Amazonian countries have established a regional Amazon Hydrological Network (AHN) and a Water Quality Network (WQN) to provide a systemic view of the Amazon basin and its flows of water, pollutants and sediments. In 2025, the eight countries engaged in a joint effort to improve the coherence of measurement and water monitoring across the basin, supported by the consolidation of four regional protocols by the Specialized Technical Panel (STP) of the ACTO Amazon Network of Water Authorities (RADA/ACTO). The protocols concern the installation and operation of hydrological monitoring stations, field analyses and sample collection, data processing and publication, and the flow and publication of AHN and WQN data.

3.3.4 Cooperation on stakeholder engagement

Successful source-to-sea management requires the active engagement of stakeholders across sectors and along the source-to-sea continuum. Enabling conditions are needed to facilitate this engagement. These include an enabling national framework; raising awareness of water, environmental and health issues; identifying a broad group of stakeholders; setting up a balanced coordination mechanism; and including provisions such as costs, time and human resources (UNECE, 2013b).

Raising awareness among upstream stakeholders of their impact on downstream environments, as well as of benefits that move upstream from downstream, can strengthen commitments to take action that preserves those benefits. In addition to institutionalized participatory processes and engagement in planning and management activities, stakeholders can be invited to participate in activities such as joint events, awareness-raising campaigns or common field trips that follow the river to the sea. These activities then become the foundation for collaborative action with stakeholders taking co-responsibility for implementation of action plans that include both individual and joint actions.

Article 16 of the Water Convention concerning public information aims to ensure that the public have access to information related to the condition of transboundary waters and measures taken, including on water quality management, pollution control and sustainable water use. In the third reporting exercise on SDG indicator 6.5.2 on transboundary water cooperation in 2023–2024, 82 countries reported that stakeholder engagement was one of the main factors in cooperating on transboundary waters, while 94 countries reported that better knowledge and understanding was a main factor (UNECE, UNESCO and UN-Water (2024)).

The SADC Protocol on Shared Watercourses, notably Article 5 (Institutional Framework for Implementation including Stakeholder Involvement), underscores the importance of stakeholder engagement in managing shared watercourses. It calls for active participation from upstream and downstream communities, local governments and other relevant sectors in water management decisions.

Article 9 of the OSPAR Convention on access to information involves a broad range of stakeholders, including the fishing industry, environmental NGOs and local governments, in the development and implementation of marine protection measures. This broad engagement ensures that both upstream and downstream stakeholders are included in efforts to reduce land-based pollution impacting the marine environment.

Article 15 of the Barcelona Convention on public participation in decision-making emphasizes the importance of engaging with stakeholders, including any sector, and environmental organizations, to ensure sustainable management of the Mediterranean Sea and its coastal areas. It mandates public consultation on plans related to pollution reduction, marine conservation and sustainable coastal development.

While not a transboundary basin, the Tamar Catchment Partnership (TCP) is an example of a multi-stakeholder platform that works to better understand the impacts of human activities in the river catchment, coast and sea, and to manage these activities in a manner that reduces negative impacts (Case Study 7).

The five-year Integrated Watershed to Reef Management of the Mesoamerican Reef Ecoregion (MAR2R) project, implemented in Belize, Guatemala, Honduras and Mexico, successfully engaged with stakeholders from the highest political level to local communities (Case Study 16).

16 Case Study
Alliances and coordination with multiple stakeholders for social, environmental and economic benefits in the Mesoamerican Reef Ecoregion

Context

The five-year project "Integrated Watershed to Reef Management of the Mesoamerican Reef Ecoregion" (MAR2R) was implemented in Belize, Guatemala, Honduras and Mexico and funded by the GEF. MAR2R was developed under the ridge-to-reef (R2R) approach, which recognizes "that everything that happens in the upper, middle and lower part of the watershed reaches the sea". The project involved multiple actors from the political level to affected communities. This allowed for the strengthening of institutional frameworks as policy and planning instruments to improve forms of governance, watershed management, coral reef restoration and demonstrations of the R2R approach, and the development of actions that helped communities improve their living conditions and income. The project was accompanied by a monitoring and evaluation strategy that ensured transparency and effectiveness in implementation to meet objectives.

Results

The project has had an impact on the conservation and sustainable use of shared freshwater, marine and coastal ecosystems in the transboundary ecoregion of the Mesoamerican Reef, while also generating economic benefits and fostering sustainable livelihoods in the communities covered by the intervention. MAR2R successfully engaged with stakeholders at different levels of participation and geography to achieve its objective, demonstrating the effectiveness of the participation and collaboration strategy, while strengthening regional governance. The involvement of stakeholders was encouraged by adapting the concept of National Intersectoral Committees to suit the circumstances of each country.

Stakeholder participation

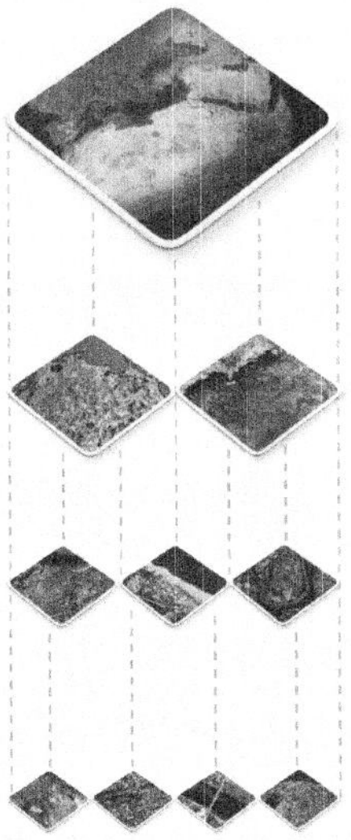

Regional level

Ministerial councils such as tourism, agriculture and fisheries, strengthening governance and promoting the watershed to reef (R2R) approach.

Intersectoral agendas between agriculture-environment and tourism-environment, extending the watershed approach to the reef to productive areas.

National level

Ministries of environment of the countries of the Mesoamerican Reef Ecoregion, emphasizing the development of policies related to integrated water and coastal marine resource management; and natural resources with watershed to reef (R2R) approach.

Sub-national level

Watershed governance platforms, private sector organizations and dialogue platforms for coral reef, mangrove and tourism management.

Local level

Civil society, communities, Indigenous Peoples and private sector companies, implementing demonstrative projects and advocating for better water management.

Lessons learned

It is essential to transform the perceptions of concerned actors from mere executors to active partners. This amplifies participation, investment and international cooperation. It is also important to strengthen the implementation capacity of regional organizations such as the Central American Commission on Environment and Development (CCAD), which can facilitate political agreements on shared resources and the implementation of regional policy instruments. The feasibility of collaborating with private sector trade associations was demonstrated, as long as the project refrained from assuming responsibilities that fall within the voluntary scope of the companies. Additionally, the assumption that organizations involved in environmental work have all the necessary capabilities hinders the strengthening of gaps in relevant areas. Instead, such gaps should be considered as opportunities for improving and strengthening organizational capabilities.

The MAR2R project achieved significant progress in regional planning, policy development and integrated management of freshwater and coastal marine resources, using a "basin to reef" approach tailored to the circumstances of each country, together with a regional perspective. The development of a second phase of MAR2R will foster public-private participation in and financing for implementation of the Strategic Action Plan (SAP). This will enable environmental benefits to be scaled up to national, regional and global levels. Overall, the experience of the MAR2R project provides a valuable model of governance and stakeholder participation, one that other initiatives can apply to the conservation and sustainable use of freshwater, marine and coastal regions in Central America and beyond.

Source: Case Study provided by Raúl Ernesto Artiga Colato, Sistema de la Integración Centroamericana (SICA), 2025.

3.3.5 Cooperation on planning instruments

Source-to-sea considerations can be incorporated into a wide variety of planning instruments, including river basin management plans, integrated coastal management plans, sea action plans, maritime spatial plans, climate change adaptation strategies, flood risk management plans, drought management plans, programmes of marine measures, strategic environmental assessments and other practical tools.

Where management plans exist or are in development (e.g. river basin management plans, coastal zone plans or marine spatial plans), source-to-sea management can strengthen the consideration of linkages between segment-specific plans and other components in the source-to-sea system. Doing so can assist the assessment of impacts – positive or negative – and the evaluation of trade-offs that might not be recognized when planning is focused on individual segments. In this way, source-to-sea management can contribute to improved outcomes and long-term sustainability. For example, the Binational Commission of the Sixaola River brings together Costa Rica and Panama for the joint management of water resources from upper areas of the basin to the sea and allows for a collaborative response at the binational level, guaranteeing the sustainability and biodiversity of the basin. In 2024, the Ministry of Environment and Energy of Costa Rica and the Ministry of Environment of Panama endorsed the Strategic Action Programme of the Binational Sixaola River Basin. (Case Study 9).

Sustainable development of the Rhine River Basin is supported by the International Commission for the Hydrology of the Rhine basin (CHR) (Case Study 1). Scientific institutes of the Rhine riparian states develop and disseminate knowledge and tools in the field of hydrology for the entire Rhine catchment area to support decision-making. In the Great Lakes Basin, the states and provinces work together on a range of issues, including implementing water management programmes; collecting comprehensive and comparable water use data; implementing water conservation and efficiency programmes; identifying regional scientific research priorities; assessing the cumulative impacts of water withdrawals, diversions and consumptive uses, as well as the anticipated impacts of climate change on the water budget; and engaging in joint decision-making on proposed water diversions that are of basin-wide interest (Case Study 5). In both cases, improved basin planning and management has reduced impacts on coastal and marine environments.

The co-chaired Joint HELCOM-VASAB Maritime Spatial Planning Working Group (HELCOM-VASAB MSP WG) was launched in October 2010 by HELCOM and the Vision and Strategies around the Baltic Sea (VASAB) Committee on Spatial Planning and Development of the Baltic Sea Region (CSPD/BSR). The Working Group was established to ensure cooperation among the Baltic Sea Region countries for coherent regional Maritime Spatial Planning (MSP) processes in the Baltic Sea. The Working Group provides a forum for regional, transboundary and cross-sectoral dialogue on Integrated Coastal Management and Maritime Spatial Planning enabling a common approach and facilitating coherent Maritime Spatial Planning across borders. It assists in implementing actions decided in the VASAB "Long Term Perspective for the Territorial Development of the Baltic Sea Region" and in MSP-related actions of the HELCOM Baltic Sea Action Plan. It provides Baltic Sea Region countries with a platform to discuss links between relevant international agreements, EU legal instruments and policies, and Baltic Sea Region and national policies.

In accordance with both regional and international conventions, the development and implementation of sectoral plans typically begins at the regional level, then subsequently adapted and operationalized at the national level through the formulation of national action plans. Within this framework, source-to-sea management can play a crucial role at the regional level by honing-in on specific sectors, facilitating the alignment of objectives, promoting stakeholder cooperation and supporting the formulation of joint regional and national action plans that ensure coherence across borders (Box 4).

Box 4

Operationalizing source-to-sea management through regional sectoral plans under the Barcelona Convention, Europe and MENA

Under the Barcelona Convention, and pursuant to Article 15 of the Protocol for the Protection of the Mediterranean Sea against Pollution from Land-Based Sources and Activities (LBS Protocol), Contracting Parties commit to the development of several policy instruments. These include regional plans focused on addressing pollution from key sectors such as municipal wastewater, sewage sludge, agriculture, aquaculture, marine litter and stormwater. These plans were formally adopted in 2021 and 2023, following in-depth consultations and sectoral dialogues among the Contracting Parties.

Key regional plans adopted under LBS Protocol Article 15

Regional Plan on Urban Wastewater Treatment (IG 25/11). The primary aim of this plan is to safeguard coastal and marine environments, as well as human health, from the harmful effects of both direct and indirect discharges of untreated or inadequately treated wastewater. The plan addresses key concerns such as oxygen depletion and eutrophication, and promotes water and energy efficiency in wastewater management, as outlined in Article II.

Regional Plan on Sewage Sludge Management (IG 25/11). Complementing the wastewater plan, this plan seeks to maximize the reuse of valuable substances and exploit the energy potential of sewage sludge. It also prioritizes the protection of human health and the environment by minimizing the negative impacts associated with sewage sludge management in accordance with Article II.

Regional Plan on Marine Litter (IG 25/9). This plan is designed to prevent and significantly reduce marine litter pollution across the Mediterranean Sea. It aims to mitigate the impact of marine litter on ecosystem services, habitats and marine species – especially endangered species – as well as to protect public health and safety and reduce the associated socioeconomic costs, as outlined in Article IV.

Regional Plan on Agriculture (IG 26/6). This plan focuses on reducing and preventing pollution arising from the use of fertilizers, pesticides and agricultural waste. It also supports practices that contribute to sustainable agriculture, helping to align agricultural development with environmental protection goals.

Regional Plan on Aquaculture (IG 26/7). Developed under the same legal framework, this plan ensures that aquaculture activities are conducted in a sustainable manner. It emphasizes pollution prevention and aims to minimize negative ecological impacts, ensuring that the aquaculture sector is managed responsibly.

Source: Erol Cavus, UNEP-MAP, 2025.

The Mediterranean Sea Programme (MedProgramme) "Enhancing Environmental Security" (2020–2026) is a six-year multifocal and cross-sectoral initiative funded by the GEF and implemented under the coordination of the Barcelona Convention. Ten Mediterranean countries – Albania, Algeria, Bosnia and Herzegovina, Egypt, Lebanon, Libya, Montenegro, Morocco, Tunisia and Türkiye – participate in the MedProgramme. Two national projects, one in Lebanon (Case Study 8) developing a Damour Integrated Management Plan and one in Morocco developing a Coastal Zone Management Plan (Case Study 17), address unique challenges and provide tailored solutions aligned with source-to-sea management. The MedProgramme supports the ratification, adoption and development of legal tools that promote innovative approaches such as source-to-sea management, IWRM, ecosystem-based approaches and the WEFE Nexus.

In the Tangiers-Tetouan-AlHoceima (TTA) region of Morocco, a multi-stakeholder Regional Coastal Governance Committee coordinated by the Directorate for the Environment has been established. The Committee identified priority actions to be taken and strengthened source-to-sea coordination to benefit coastal and marine ecosystems. Source-to-sea management is integrated into a broader TTA Regional Development Plan. Planned next steps are to identify potential areas for future improvement and scaling-up source-to-sea management practices and efforts (see Case Study 17).

17 Case Study

Regional governance for source-to-sea management in the Tangiers-Tetouan-AlHoceima region, Morocco

Context

In the Moroccan region of Tangiers-Tetouan-AlHoceima (TTA), activities undertaken by the MedProgramme have included developing an Integrated Coastal Zone Management (ICZM) plan through an inclusive participatory process, conducting multiple assessments including Water-Energy-Food-Environment (WEFE) Nexus assessment, coastal climate risk assessment, studies on the Rhiss-Nekkor coastal aquifer and establishing a multi-stakeholder Coastal Governance Committee coordinated by the Regional Directorate for the Environment. The MedProgramme identified and prioritized climate risk mitigation measures such as infrastructure for coastal protection, reforestation projects to prevent soil erosion and preserve groundwater recharge, upgraded water treatment plants with innovative climate-friendly technologies, and low-carbon desalination plants to address precipitation variability. The project also integrated source-to-sea management into the broader TTA Regional Development Plan, facilitated capacity-building for monitoring the impact of water discharges on coastal and marine water quality and associated ecosystems, strived to control non-indigenous species through developing dedicated fisheries and supported the adoption of sustainable fisheries practices harmonized with ecosystem protection to strengthen the socioeconomic resilience of coastal communities.

Challenges

Key challenges that needed to be addressed were the fragmented governance framework for water management, ensuring coherent implementation along the source-to-sea continuum across different sectors and institutions, limited funding to support pilot implementation of the ICZM plan and the WEFE Nexus recommendations, and limited legal provisions for coordinating coastal zone, freshwater and marine ecosystem management.

Activities

The project created a Regional Coastal Governance Committee to pilot an innovative coastal resource management governance framework, established a Regional Steering Committee to assist the Regional Council in consolidating integrated approaches for the management of natural resources, developed a common spatial framework encompassing all segments of the water continuum, coordinated planning tools across different water continuum segments to provide consistent management, and balanced different users' needs in the context of climate change and variability.

Lessons learned

The project demonstrated that the source-to-sea approach effectively supports spatially integrated water management, unifies management across water continuum segments and ensures balanced resource access. Successful implementation requires supportive legal and institutional frameworks for coordinated decision-making, ongoing dialogue between science and policy (Science-Policy Interface), shared understanding of environmental and sustainability issues, a culture of integration for common objectives and investment plans to support financially sustainable pilot projects.

Source: Case Study provided by Mohamad Kayyal and Alessandro Candeloro, UNEP-MAP, 2025.

The importance of healthy freshwater ecosystems has been recognized in key international development frameworks. However, freshwater ecosystems are still undervalued, under-funded and overlooked, resulting in one-quarter of freshwater fauna being threatened with extinction. To change this trend a group of six concerned governments (Colombia, Democratic Republic of Congo, Ecuador, Gabon, Mexico and Zambia) mobilized to launch the Freshwater Challenge (FWC) at the UN 2023 Water Conference. This country-led initiative has been submitted as a commitment to the Water Action Agenda (UN Water), a Water Outcome of COP28, part of the COP29 declaration on water for climate action, and a flagship initiative under the Riyadh Action Agenda, and sits under the auspices of the UN Decade of Ecosystem Restoration.

The Freshwater Challenge aims to restore 300,000 km of degraded rivers and 350 million hectares of degraded wetlands by 2030 as well as secure the protection of freshwater ecosystems important for biodiversity and ecosystem services. Restoration needs to happen along the length of rivers and is dependent on effective transboundary cooperation within a source-to-sea framing. This ambitious goal is linked directly to achieving the 30x30 targets in the Kunming-Montreal Global Biodiversity Framework.

By March 2025, 49 national governments and the European Union had joined the Challenge and committed to set and thereafter implement quantifiable targets in their national plans to restore and conserve freshwater ecosystems. Those targets can be incorporated into relevant national policies, plans and strategies, including plans under international agreements and agendas, such as National Biodiversity Strategies and Action Plans (NBSAPs), Nationally Determined Contributions (NDCs) and National Adaptation Plans (NAPs), Climate Action Pathway for Water, Land Degradation Neutrality (LDN), sustainable development plans and National Plans for Networks of Wetlands, among others.

Non-state actors and private sector entities can also join the Freshwater Challenge to accelerate the delivery of national targets supporting countries to operationalize their FWC commitments in national plans. The FWC is designed to harness, connect to and build upon existing platforms and complementary initiatives to substantiate existing ambitions designed to safeguard freshwater habitats and species. Through a coordinated approach, the Challenge aims to influence the global agenda, improve upstream and downstream connectivity as well as river flow, and increase resources available for these efforts. Through its efficient and agile design, the Challenge responds to innovative financing solutions, thereby increasing investment in freshwater ecosystem restoration and conservation between sectors and across borders.

© Wasunan / Adobe Stock

Chapter 4

Practical steps toward transboundary source-to-sea management

Key Takeaways:

- The source-to-sea approach offers a structured six-step process to tackle source-to-sea challenges – from problem definition to measuring outcomes and capturing lessons learned. It considers social, environmental and economic linkages and impacts across source-to-sea segments to identify and prioritize system-wide issues.

- The source-to-sea approach consists of six steps:
  1. Characterize the biophysical system;
  2. Engage with key stakeholders;
  3. Diagnose the policy and governance system;
  4. Design the way forward;
  5. Co-create an action plan and foster an enabling environment;
  6. Monitoring, evaluation, learning and adaptation.

- Shared understanding of the six key flows (water, biota, sediment, pollution, materials and ecosystem services), stakeholder roles and governance dynamics is a prerequisite for embedding source-to-sea management in transboundary cooperation.

- Each regional, national and/or local context has a source-to-sea management readiness level and the activities undertaken need to reflect the current level.

- Identifying system-wide priorities focuses transboundary cooperation on shared objectives, but success depends on implementation of a coordinated action plan covering joint and individual efforts. The source-to-sea approach ensures that actions taken at every stage of transboundary cooperation contribute to sustainability, resilience and equitable benefit-sharing from source to sea.

- Monitoring progress, sharing insights and adaptively managing implementation are central to advancing source-to-sea outcomes. Through collaboration, prioritization and learning, the source-to-sea approach supports lasting improvements in ecological integrity, social equity and economic resilience.

At a glance: case studies in this chapter

4.1 Steps of the source-to-sea approach

The source-to-sea approach provides a structured process to address source-to-sea challenges ranging from initial problem definition to results measurement and lessons learned (Mathews and others, 2019). It generally follows the project cycle and aligns with steps taken in transboundary cooperation and under the Water Convention. The approach is flexible and can be adapted to the local context. Application can start with any step relevant to the situation. While the steps follow a logical order, they are closely interlinked. Iterations of previous steps may be developed as new understanding arises in subsequent steps. The intended outcome of the approach is to identify appropriate courses of action that will result in economic, social and environmental benefits for the source-to-sea system as a whole.

The approach includes six steps (Figure 4) through which linkages between source-to-sea segments, activities and impacts are considered in order to identify and prioritize issues to be addressed across the whole system. The approach commences with understanding the pressures and drivers of altered key flows and identifying the relevant sectors. The next stages involve selecting an appropriate scale of intervention, engaging with stakeholders (both upstream and downstream) and gaining a thorough understanding of the governance context. This sets the basis for identifying intervention strategies and will guide planning and implementation. Monitoring and adaptive management round out the process and can be used to ensure continuous improvement towards long-term outcomes.

Figure 4 **Six steps for taking practical action towards transboundary source-to-sea management**

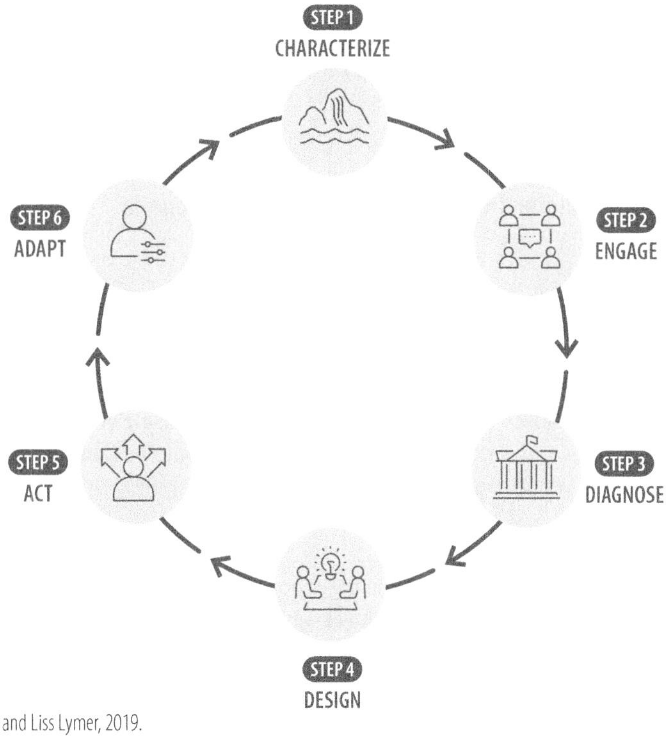

Source: Mathews, Tenberg, Sjödin and Liss Lymer, 2019.

Developing a shared understanding of source-to-sea linkages is a crucial step for cooperation between States in setting goals and objectives at the source-to-sea system level. This can draw on existing or new research, studies, models and monitoring, stakeholder surveys, citizen science and other sources, and incorporate knowledge from local communities and Indigenous Peoples. As source-to-sea systems are complex and dynamic, complete knowledge of all environmental, social and economic interrelationships and impacts to one or more key flows across different segments is unrealistic. Instead, the goal is to draw on existing knowledge to ascertain how these elements interact with each other. The focus is to develop a holistic understanding of the source-to-sea system, which continues to evolve over time as new information emerges.

4.1.1 Step 1: Characterize the biophysical system

Characterization of the biophysical system begins with identifying the undisturbed or natural ranges of variation of one or more of the key flows – water and sediments, native biota populations and their habitat uses, water quality, materials and geomorphology, and ecosystem services – within and across the segments of the source-to-sea system. The purpose is to develop a clearer understanding of the relationships between segments of the source-to-sea system, and to select one or more priority key flows and the system boundary as the focus of future interventions.

Characterizing key flows in their undisturbed state, to the extent possible, provides a baseline for the analysis of alterations. Assessment of how these flows have been altered, and the sources and impacts of those alterations, can enhance understanding of the linkages between environmental changes and social and economic values across source-to-sea segments. It is important to bear in mind that the scope of analysis for Steps 2–6 is determined by the priority key flow(s) and the system boundary defined in Step 1. This is done purposefully to help maintain a focused approach.

Knowing the scope, scale and severity of impacts, their sources and the affected locations, is crucial to prioritizing and targeting responses. Any analysis of sources and impacts should consider the source-to-sea system as a whole – momentarily blind to national borders – in order to create a holistic overview that can be used to prioritize actions at the system level. In transboundary systems, cooperation among States will be necessary to determine priorities for the source-to-sea system, which may differ from national priorities. Similarly, the analysis should be broadly inclusive of sectors, reflecting in full the activities affecting the health of the source-to-sea system.

A wide range of local knowledge and existing and modelled data can be used to characterize key flows. In Indonesia, this combination of best available data and modelling was used to create a spatially and temporally variable representation of the generation, transport and fate of plastic waste from source to sea (Case Study 18).

18 Case Study
Characterizing plastic waste pollution from its source to the sea in Indonesia

Context

Despite political commitments and ambitious targets to reduce plastic pollution by countries like Indonesia, knowledge of the sources and quantities of mismanaged plastic waste and leakage into the environment are still limited. In 2019, the World Bank financed a national study to estimate plastic waste leakage from land-based sources in Indonesia and establish a robust but detailed national baseline of plastic litter input into the sea. This study was led by Deltares and was closely supported by the Indonesian Coordinating Ministry for Maritime Affairs and Investment, which had responsibility for marine litter policies in Indonesia.

Activities

The assessment was national in scope but involved the collection of detailed information from a broad range of Indonesian data sources at the local level. Material flow analyses were performed in all the main river basins and solid waste management practices were assessed at each of the 514 kabupaten/kota (regencies). This approach allowed for insights at both the national and local scale. These data were then integrated with data on national rainfall, topography and river flow averages, and movements of plastic waste generated on land into waterways. Modelling was then conducted using state-of-the-art numerical models and real-time hydrological data.

This process generated a spatially and temporally variable representation of the generation, transport and fate of plastic waste from source to sea, providing information on waste generation and leakage within each river basin. Catchments were ranked based on their contributions to plastic inputs into the sea and detailed summaries for the main regencies and regions were provided on the amounts and sources of plastic waste leakage. Recommendations for policy interventions and actions were tailored to these outcomes.

Results

This detailed information can help the Indonesian government tackle plastic pollution by pinpointing where and which interventions are most needed to prevent plastic leakage at its source, as well as providing a realistic baseline against which progress in preventing plastic inputs into the sea can be tracked. The assessment is thus a valuable tool in source-to-sea plastic pollution prevention.

Source: Case Study provided by Joana Mira Veiga, Deltares, 2025.

Hazardous substances pollution is a significant water management issue. Addressing this issue requires basin-wide pollution assessments and harmonized abatement measures. To help achieve the goal of a "toxic-free" Danube River Basin and, consequently, prevent hazardous substances from entering the Black Sea, the Danube Hazard m^{3}c project worked to reduce the data and knowledge gap on chemical pollution, thereby supporting the development of policies, regulatory frameworks and measures that reduce chemical pollution (Case Study 19).

19 Case Study

Narrowing the data and knowledge gap on hazardous chemicals to achieve a "toxic-free" Danube River Basin

Context

The project "Danube Hazard m^{3}c", financed from the Interreg Danube Transnational Programme, aimed to improve knowledge and understanding of the status of water pollution from hazardous substances in the Danube River Basin. The project brought together 10 partners from 8 countries representing research institutes, environmental agencies, laboratories, ministries and international organizations. Project partners implemented the three core elements of water governance – measuring, modelling and management – in innovative ways, complemented by capacity building.

Activities

The project delivered a comprehensive inventory of chemical concentrations by integrating existing measurement data and the results of targeted monitoring campaigns. For the first time ever, a chemical emission model was produced by Deltares for the entire basin, developed by updating a previous Danube water quality model and upscaling the findings of detailed model investigations focused on pilot catchments. A technical manual, policy guidance and a measurement catalogue were also provided to improve the management of hazardous substance pollution, and several training events took place to familiarize national experts and stakeholders with these tools.

Challenges

Hazardous substances pollution represented a significant water management issue in the Danube River Basin, and one that required basin-wide pollution assessments and harmonized abatement measures – approaches that imply source-to-sea management. The main issue concerned the persistent presence of chemicals in the aquatic environment, undermining the "good" status of surface water bodies. To achieve a "toxic-free" Danube River Basin, further steps needed to be taken. The first step was to narrow the information gap related to chemical pollution and enhance current knowledge of monitoring and chemical emissions. In tandem, the water management sector (both freshwater and marine) needed to adapt its policies, regulatory frameworks and measures to reflect the latest EU requirements on reducing chemical pollution. A further challenge was that in some Danube countries, substantial lack of institutional capacity and insufficient intersectoral dialogue hindered the establishment of efficient management. There was also a need for adaptation measures to counteract the rapidly increasing pressures of climate change impacts on water quality management.

Results

The Danube Hazard m^{3}c project increased available knowledge and narrowed data gaps with regard to monitoring, modelling and management of chemical pollution in the Danube River Basin. In addition, capacity building improved capabilities to share information while enhancing the skills of water professionals. Importantly, a science-policy interface was established by linking scientific investigations with policy recommendations. The resulting technical knowledge, practical tools and guidance documents, jointly developed by the project partners, will be highly useful for Danube countries in tackling hazardous substance pollution at national and transboundary levels.

Danube Hazard m^{3}c implemented a holistic approach to collect data, develop models and provide guidance for managing hazardous substances pollution in the Danube River Basin. Tools developed for source-to-sea management at the basin-wide level were supported by local investigations in pilot catchments. The project also established an international partnership consisting of key actors and close cross-border cooperation based on mutual trust, knowledge sharing and transparency. By working together, the project substantially improved knowledge and understanding of hazardous substances pollution in the basin, paving the way towards a toxic-free Danube River Basin.

Lessons learned

Current complex changes in EU water policy and external pressures require a new transnational action plan to manage chemicals in the Danube River Basin. To reflect these challenges, the transnational database and model, despite being valuable tools, need to be adapted to meet present and future needs of the ICPDR and Danube countries. Such enhanced tools would offer great opportunities to update chemical pressure assessment processes in the Danube River Basin, support transnational measure implementation, and prioritize and coordinate actions and measures, taking into account different national initial conditions, resources and possibilities. Danube countries would also be provided with operational data management and pollution assessment tools to support practical water management at the national level.

Source: Case Study provided by Adam Kovacs, ICPDR, and Jos van Gils, Deltares, 2025.

4.1.2 Step 2: Engage with key stakeholders

Understanding the social and economic aspects of the source-to-sea system is essential for the success of source-to-sea management. States have the international obligation to use a transboundary watercourse in a reasonable and equitable manner, and in doing so, to balance the different interests and demands of stakeholders in the utilization of transboundary watercourses. Characterization of the biophysical system, including the identification of alterations to key flows and their sources and impacts, lays the foundation for identifying stakeholders whose engagement is essential for transboundary cooperation. Stakeholders can be mapped in relation to their social and economic dependencies on key flows, and how they are impacted by these alterations, or if their activities alter key flows.

It is important to involve in this process Indigenous and ancestral peoples, marginalized and vulnerable communities, women and youth, and other groups commonly excluded from decision-making. Better understanding of the interests of stakeholders relative to the segments of the source-to-sea system will help guide engagement activities. An assessment of the relationships between actors across the source-to-sea system both vertically (from local to national or transnational levels) and horizontally (across all public institutions, local communities or bilateral commissions) will also enrich this process.

Mapping and engagement of stakeholders will need to account for varying levels of influence related to specific issues and the power dynamics between stakeholders both vertically and horizontally. How a stakeholder will be mapped is dependent upon specific key flows; for example, a stakeholder's economic interests may be dependent on there being little or no alteration to a particular key flow, while their activities alter another key flow. Similarly, the mapping of stakeholders and the power dynamics may vary across segments of the source-to-sea system.

When mapping stakeholders in a transboundary system, the analysis of stakeholders must be conducted in relation to source-to-sea segments, both within the national context and at the source-to-sea system level. The geopolitical dynamics between States will need to be considered in addition to stakeholder relationships within the source-to-sea system. In this regard, Article 16 of the Water Convention states that information on the conditions of transboundary waters and measures regarding transboundary impact shall be made available to the public, allowing for the engagement and participation of stakeholders and transparency in the utilization of transboundary waters.

Table 5 presents a simple categorization of stakeholders that can be useful in stakeholder mapping and the development of an engagement strategy. Primary and external stakeholders can be important for strengthening necessary political will to invest in source-to-sea management. Raising their awareness of source-to-sea linkages and the benefits that can accrue from source-to-sea management can help mobilize them as advocates for this approach. Incentives to promote change among target stakeholders may be provided by enabling and supporting stakeholders, who are best positioned to create the necessary enabling environment to achieve changes in behaviours and practices among target stakeholders. To this end, they may employ legal, financial and/or technical support. It is possible for an individual or a group to fall into one or more stakeholder mapping categories at the source-to-sea system level.

Developing an engagement strategy based on these categories will support the inclusive participation of relevant stakeholders, reflecting their roles in the process of developing source-to-sea management. For example, river basin and regional seas commissions create an enabling environment for stakeholder engagement and facilitate the interface between States. Mapping stakeholders in this way supports the design of appropriate governance schemes and can help in the identification of actions needed to address source-to-sea challenges.

Table 5 Categories for stakeholder mapping

Stakeholder category	Description
Primary	Primary stakeholders may be dependent on ecosystem services, are affected by alterations in key flows, and are beneficiaries of the environmental, social and/or economic improvements achieved through source-to-sea management.
Target	Target stakeholders are actors or sectors whose activities are contributing to the alteration of key flows and whose behaviour will need to change to realize benefits from source-to-sea management.
Enabling	Enabling stakeholders provide the enabling conditions for behaviour changes to occur and benefits to be sustained over time.
Supporting	Supporting stakeholders include development partners, financiers and others whose goals are aligned with and can support the realization of benefits from source-to-sea management.
External	External stakeholders are individuals or groups outside the source-to-sea system who share an interest in one or more of the benefits arising from source-to-sea management.

Source: Mathews, Tenberg, Sjödin and Liss Lymer, 2019.

A Source-to-Sea Learning Tour was conducted in the Orange-Senqu River Basin to enhance stakeholder knowledge about the challenges, opportunities and infrastructure pertaining to water usage. Participants included both public and private sector representatives from ORASECOM's four State Parties (Case Study 20).

20 Case Study

Raising stakeholder awareness through a Source-to-Sea Learning Tour of the Orange-Senqu River Basin

Context

The Orange-Senqu River Basin is located in southern Africa and spans Lesotho, South Africa, Botswana and Namibia, emptying into the Atlantic Ocean. It plays a crucial role in regional water resource management. In response to a request submitted by member states of the Orange-Senqu River Commission (ORASECOM), who wished to acquire a comprehensive understanding of the entire basin's physical geography, and the range of projects and stakeholders involved, the ORASECOM Secretariat organized a Source-to-Sea Learning Tour. The tour was supported by the UNDP-GEF Strategic Action Plan (SAP) Implementation Project, and was attended by representatives from ORASECOM's four State Parties, including from the public and private sectors.

The objective of the Learning Tour was to enhance knowledge about the challenges, opportunities and infrastructure pertaining to water usage, while fostering stakeholder engagement and capacity building. The overall aim was to assist in giving sound, relevant and pertinent advice to stakeholders and State Parties, taking into account the impact of developments on water management, water use, pollution control and biota.

Activities

The Tour included visits to key infrastructure such as the upstream Katse and Mohale Dams in Lesotho, the midstream Vanderkloof and Gariep Dams in South Africa, and the downstream Hardap and Neckartal Dams in Namibia. Meetings and discussions were held with stakeholders, including dam operators, water supply institutions, energy generation companies, agricultural irrigation farms/schemes and mining companies.

Challenges

Identified challenges, opportunities and emerging threats related to: water use, existing and planned infrastructure; over-utilization of water resources; environmental issues from over-abstraction for industrial, agricultural and mining uses in the basin; polluted water resources, especially in the Gauteng Province of South Africa, which accounts for the highest level of utilization, and from the textile industry in Lesotho; expansion of agricultural land in the upper and lower Orange; and sedimentation in the upper reaches in Lesotho. Related recommendations address equitable allocation, the maintenance of environmental flows, drafting of a river mouth management plan and enforcement of water quality monitoring.

Issues identified during the Learning Tour included climate change, linkages between freshwater and marine ecosystems, impacts of low river levels on irrigation canals, downstream flooding and damage to water management infrastructure, water security and quality issues including high levels of salinity, eutrophication and plastics pollution. Additional identified challenges included inadequate financial and technical resources hindering sustainability efforts and monitoring of ecosystem health, and gaps in hydrological, water quality and biodiversity data affecting decision-making and planning.

Results

The Learning Tour enhanced interaction among various stakeholders in water resources management and increased understanding of the interconnectedness of upstream and downstream water infrastructures. Key issues highlighted included the need for comprehensive Environmental Flow Requirements (EFRs) to support ecosystem health and the urgency of solid waste management issues in the basin's upper reaches. The Tour increased awareness of the importance of maintaining ecosystem health, improved understanding of national and regional water resource management policies as well as the source-to-sea concept, empowered local communities to engage actively in resource protection, and strengthened institutional cooperation aimed at promoting sustainable governance of transboundary water resources.

Through the Source-to-Sea Learning Tour, ORASECOM effectively inducted its stakeholders into the complexities of the Orange-Senqu River Basin. The tour focused particularly on the impacts of river development on the quantity and quality of freshwater inputs into the marine and coastal environments, and the need to jointly manage the two ecosystems. This initiative created a foundation for ongoing collaboration among the four member States, while also laying the groundwork for future studies and actions to sustain water resources management in the region.

Namibia led the way in southern Africa by ratifying the 1992 UN Water Convention, followed by Zambia and Zimbabwe. Botswana and Lesotho have started the process of ratification. The overall framework for integrated and cooperative water resource management continues to evolve, ensuring the long-term sustainability of the Orange-Senqu River Basin and the maintenance of the entire source-to-sea system.

Lessons learned

The information generated by the Learning Tour will feed into ORASECOM's reviews of existing plans including the Strategic Action Plan (SAP) and foreseen National Action Plans (NAPs). It will also add value to studies commissioned to determine Water Resource Quality objectives, basin-wide environmental flows, management of invasive species to prevent biodiversity loss and pollution prevention. Looking ahead, it is imperative to conduct a comprehensive source-to-sea management study to further solidify understanding of the basin's dynamics, as previously explored in a 2012 study conducted by ORASECOM and the Benguela Current Commission with the support of UNDP-GEF. It is also essential to emulate global perspectives on source-to-sea management, and encourage cooperation and commitment among member states in addressing the identified challenges and ensuring sustainable management.

Source: Case Study provided by Shishani Munyika, Ministry of Agriculture, Fisheries, Water and Land Reform of Namibia, 2025.

Gariep Dam on the Orange River. © Jurgo Van Wyk, Department of Water and Sanitation, South Africa

Action to improve source-to sea communication and coordination between stakeholders, this time around wastewater treatment, was also at the heart of an initiative in India. Recognizing that limited communication and coordination was hindering efforts to reduce nutrient loads and wastewater discharge into the Yamuna River, UNEP and the Sustainable India Trust brought together key stakeholders from the government, private sector, NGOs and academia (Case Study 21). These upstream efforts to improve wastewater treatment in New Delhi will reduce overall discharges into the Ganges River and, ultimately, the Bay of Bengal.

21 Case Study

Building bridges between policymakers, academics and non-governmental organizations for upstream management of nutrient loads and wastewater discharge into the Ganges River and the Bay of Bengal, India

Context

The National Capital Territory of Delhi, where New Delhi is located, is situated along the River Yamuna, one of the main tributaries of the River Ganges, which empties into the Bay of Bengal. Wastewater is largely untreated in this metropolitan area, home to 30 million inhabitants, due to lack of sufficient infrastructure, while outdated technologies further exacerbate the problem. As a result, nutrients from wastewater pollute surface and underground water bodies within the region leading to significant eutrophication and biodiversity loss. This poses substantial risks to the health and well-being of both people and nature, affecting particularly prominent water sources such as the lakes of Najafgarh, Sanjay and Hauz Khas, along with the Yamuna River.

Seeking to address its wastewater challenges, Delhi has set an ambitious policy target to treat over 95 per cent of the city's wastewater. A variety of activities are being carried out achieve this goal, including increased sewer pipeline connections targeted especially at informal settlements, piloted treatment techniques, enlarged treatment capacity at current plants and the construction of new ones.

Activities

To support Delhi's quest to reduce nutrient loads and wastewater discharge into the Yamuna River, UNEP and the Sustainable India Trust – a partner of the Global Wastewater Initiative (GWWI) and the Global Partnership on Nutrient Management (GPNM) – implemented an initiative focused on assessing the recovery potential of nutrients in wastewater and mapping current recovery and reuse practices. These practices were compared with available technologies, leading to the identification of appropriate technologies suitable for use in Delhi. Additionally, ecosystem health cards were created to track improvements in the state and recovery of water bodies, based on water quality parameters for nutrient loads. A stakeholder workshop organized with representatives from local and national authorities, academia and non-governmental organizations (NGOs), communicated the results of the intervention, building a shared understanding of the current situation and informing policy development. The workshop also discussed next steps with a view to stemming pollutant flows.

Challenge

A notable challenge was limited communication and coordination between stakeholders involved in addressing wastewater pollution, especially scientists and policymakers. The initiative worked to improve dialogue between stakeholders and create an enabling environment to identify a shared pathway forward, and make actionable and informed policy recommendations.

Activities

The initiative convened major stakeholders from the government, private sector, NGOs and academia to inform them about the outcomes of this intervention and provide them with policy recommendations. This approach was well received and acknowledged by the government entities working on water, such as the Delhi Jal Board, which expressed its intention to implement the recommendations. The abovementioned ecosystem health cards, developed for the initiative, highlighted the status of the lakes and tracked environmental improvements made through the intervention. They also supported policymakers in developing scientifically supported policies and updating existing ones.

Lessons learned

This case study demonstrates the vital importance of understanding the local context and the relationships between different stakeholders, when seeking to foster coordination and collaboration as part of source-to-sea action. As was the case here, this may necessitate the creation of communication channels between key actors. UNEP and the Sustainable India Trust achieved this goal by facilitating information exchanges where relevant stakeholders could learn from one other, discuss potential actions and identify innovative solutions. A key focus of this approach was bringing together public officials and scientists. As synergies are identified and trust is built, these channels can be widened and eventually formalized in governance frameworks. The outputs of this study will provide valuable inputs to the efforts of government entities like the Delhi Jal Board, the Delhi Development Authority and the Ministry of Environment, endeavouring to control wastewater, nutrient and microplastic pollution in Delhi.

Crucially, the initiative built on current and ongoing efforts in Delhi by the relevant stakeholders to address wastewater pollution. For example, the ecosystem health cards helped understand how anthropogenic activities contribute to the health of water bodies in both positive and negative ways. Additionally, the analysis of nutrient recovery technologies was used to provide recommendations to decision-makers regarding the choice of technologies best suited to the local context.

Next steps

The project team will liaise closely with relevant authorities and monitor policy developments resulting from the project. Once funding becomes available, a third phase of the project will launch, building on existing outcomes to expand the focus to include nutrient and microplastic pollution sources upstream from Delhi and examining downstream impacts of nutrients along the flow of the Yamuna River up to the Ganges River.

Source: Case Study provided by Riccardo Zennaro and Avantika Singh, UNEP, 2025.

4.1.3 Step 3: Diagnose the policy and governance system

The state of a source-to-sea system is influenced by its governance system. Any analysis of that governance system will therefore provide important background information on the context for transboundary cooperation and source-to-sea management. The diagnosis will seek to ascertain the mandates, roles and responsibilities of institutions, identify the main instruments shaping behaviours and practices that result in alterations to key flows, and understand the impacts of those alterations. These arrangements may be both formal and informal. Together, institutions and the instruments they use structure the relationships and decision-making space for source-to-sea management and transboundary cooperation.

The primary focus of the governance analysis is enabling stakeholders, often public institutions, but may also include the private sector and civil society actors. Table 6 lists some of the institutions and instruments related to governance functions that are relevant to transboundary cooperation (Jimenez and others, 2020).

Table 6 Elements of governance functions in transboundary waters

Function	Institutions, instruments and activities related to governance functions
Policy and strategy	• Regional water policies • Foreign policy • Policies governing regional economic integration • Basin, sub-basin or seas treaties, strategies or policies • Obligations under conventions and Multilateral Environmental Agreements of which a country is a party • Policies and incentives in sectors affected by water (e.g. agriculture, fisheries, transport, etc.)
Coordination	• Established technical and high-level political coordination (e.g. sharing hydro/meteorology data) • Operation of infrastructure with transboundary impacts • Joint conservation efforts • Joint management bodies such as RBOs and RSCAPs
Planning and preparedness	• Regional level (cross-sectoral) plans • Master Plans • Basin management plans • Feasibility studies and joint benefits assessments
Financing	• Joint financing of transboundary water infrastructure or management institutions • Distribution of financial risk • Development of regionally shared resources (e.g. power pools)
Management arrangements	• Organizational and institutional arrangements for shared waters (e.g. Transboundary River Commission, Transboundary River Basin Authority, Regional Seas Commission, etc.) • Joint infrastructure development, operation and management and maintenance • Mechanisms in place for coordination of joint use of infrastructure (hydropower developments, flood protection, cascading dam management) and emergency uses • Regional economic commissions or environmentally focused regional institutions
Monitoring, evaluation and learning	• Early/accident emergency warning systems • Systems monitoring in shared waters • Periodic status assessments • Outcome documents, publications, public events
Regulation	• International water laws and conventions/norms, declarations, protocols, commitments and rules governing the relations and arrangement between shared waters • Regulations (i) through transboundary agreements; and (ii) by integrating transboundary rules and national rules (e.g. Environmental Impact Assessments (EIAs))
Capacity development	• Developing consensual knowledge around shared water resource availability and quality • Developing capacities for relevant actors (RBOs, Regional seas and Large Marine Ecosystem (LME) Commissions, Coastal Management agencies, Regional Economic Commissions, CSO, Media, etc.)

Analysis of institutions can include information and discussion about their mandates relative to key flows and source-to-sea segments, geographic scale of responsibility, relevant governance functions, key instruments, level of implementation, implementation barriers and other information. Instruments may be generally applicable, providing direction to decision-making or priority setting on social, environmental and/or economic issues, or may relate specifically to a particular sector. Some instruments may be administered by one or more institutions, or outline roles and responsibilities for several tiers of governance. Many instruments that are general in nature form the foundation of subsequent legal instruments, such as legislation or regulations that are more specific in nature, bylaws which may be locally driven rules, plans that are rules-based or strategic, and funding strategies.

The governance analysis is used to identify key overlaps or gaps between the mandates of different institutions, conflicts and coordination challenges, barriers to implementation, and responsibilities relative to practices or behaviours that alter key flows. It also evaluates the power relations between institutions and their related instruments. These need to be assessed at the source-to-sea system level as well as national and subnational levels. Overlaps occur when more than one institution or instrument is managing the same source-to-sea segment, key flow or issue for similar purposes. In cases where overlaps exist (e.g. when river basin management is the responsibility of both a national authority and the transboundary basin organization), the relative mandates, functions and tasks may differ. Gaps occur when no institution or instrument addresses a segment, key flow or issue, or when there is a lack of enforcement capacity. Conflicts may occur when one or more institutions hold responsibilities over the same issue, but their mandates or objectives can negatively impact one another.

Such overlaps, gaps and conflicts can have a geographical as well as a transnational component, where the management of one segment of the source-to-sea system by an institution can result in upstream and downstream impacts; they can also occur within the same segment where different institutions are responsible for a key flow. In some cases, specific instruments or institutions may not manage all aspects of the key flow, governance instead relying on a piecemeal approach. Objectives and targets may also be poorly defined. The diagnosis of the governance system clarifies where opportunities exist to strengthen source-to-sea management and identifies challenges and barriers that need to be addressed.

Figure 5 illustrates the longitudinal profile of the uMngeni River source-to-sea continuum and the institutions and instruments that govern the river, riparian and marine ecosystems. It presents a spatial analysis of overlapping legislation at the provincial and national levels (HAV, 2019). This demonstrates clearly the need for coordination between governance actors and stakeholders if policy coherence across the continuum is to be achieved.

The Mediterranean Sea Programme (MedProgramme) found that only limited legal provision or governance mechanisms exist in many countries to support inter-institutional and cross-sectoral integrated work on freshwater and marine ecosystem management along the source-to-sea continuum. Accordingly, the MedProgramme took action to develop or establish such mechanisms either by building upon existing structures and expanding their role and mandate, or by creating new coordination structures to address specific integrated approaches and streamline sectoral outcomes as part of a wider systemic source-to-sea approach. While such mechanisms proved efficient and effective in terms of their contribution to the participatory and integrated development of required plans, their sustainability is a concern when plans are not supported by functional and coordinated financing and implementation mechanisms.

As part of the IWEco project (Case Study 22), the Integrated Management System Assessment Model (IMSAM) was developed to guide a review of legislative and institutional capacity to support Sustainable Land Management (SLM) and water resources and ecosystems management in participating states. The model was used to determine gaps which then informed priority interventions within IWEco participating states. The resulting national activities were extensive:

- development of a Methodological Framework for ICZM for Grenada, Jamaica, Saint Lucia and Saint Vincent, and the Grenadines;
- establishment of a National Water Information System for St. Kitts and Nevis;
- creation of a National Research Strategy for the Collection of Environmental Data Necessary for improved Decision-making in Antigua and Barbuda and Trinidad and Tobago;
- development of a Marine Pollution Management Strategy and Action Plan for Barbados;
- revision and production of an updated National Risk Management Plan for the Dominican Republic.

The project also supported the development and implementation of a Regional Environmental Monitoring Data Portal to assist in tracking both local and transboundary benefits from national projects.

Figure 5 Schematic of governance instruments and institutions along a source-to-sea continuum

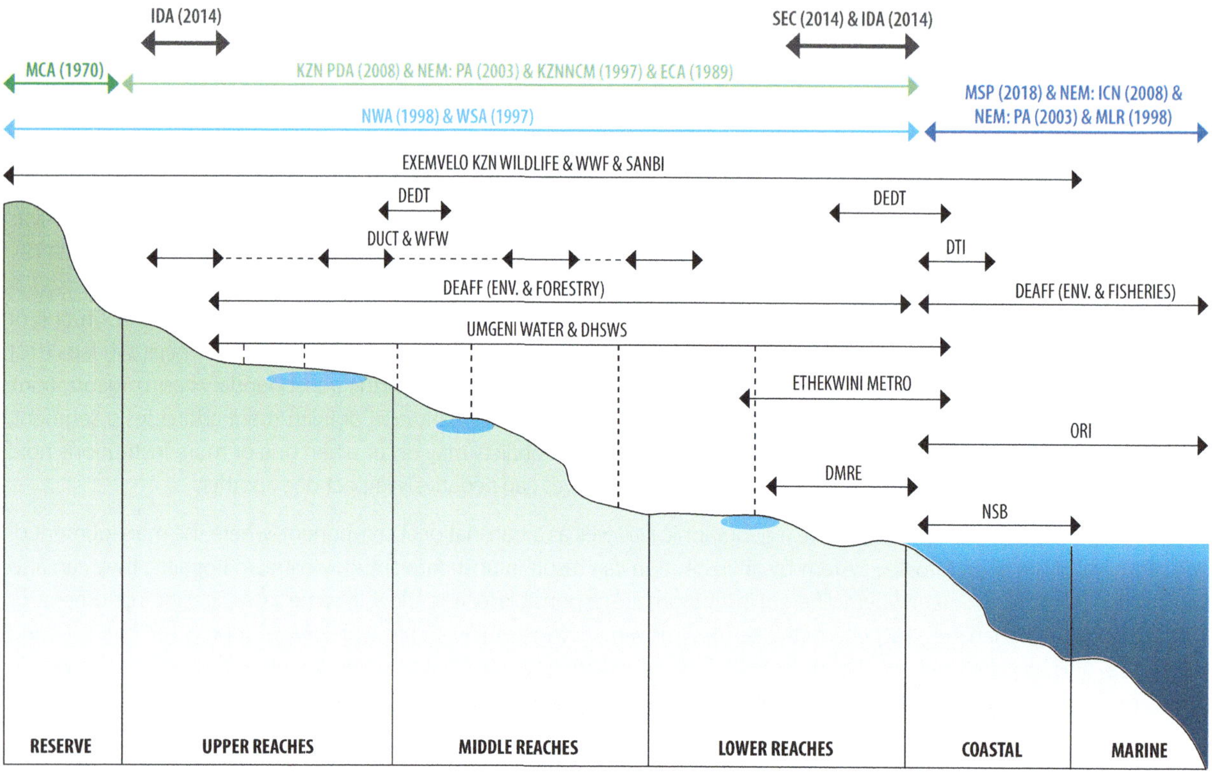

Note: Bold coloured arrows and their associated labels at the top of the figure indicate the main legislation that governs the environment. The thin black arrows and their respective labels denote the key stakeholders responsible for governance in different segments. Blue ovals represent dams and their associated reservoirs. Vertical dashed lines denote some of the monitoring points along the continuum that are governed by specific stakeholders. Horizontal dashed lines indicate that those stakeholders work in different areas along the continuum.

Acronyms: DEAFF: Department of Environmental Affairs and Forestry; DEDT: Department of Economic Development and Tourism; DHSWS: Department of Human Settlements, Water and Sanitation; DMRE: Department of Mineral Resources and Energy; DTI: Department of Trade Industry and Competition; DUCT: Duzi Umgeni Conservation Trust; ECA: Environment Conservation Act; KZN PDA: KwaZulu-Natal Planning and Development Act; KZNNCM: Kwazulu-Natal Nature Conservation Management Act; MCA: Mountain Catchment Areas Act; MLR: Marine Living Resources Act; MSP: Marine Spatial Planning; NEMBA: National Environmental Management Biodiversity Act; NWA: National Water Act; NEM:ICM: National Environmental Management: Integrated Coastal Management Act; NEM:PA: National Environmental Management: Protected Areas Act; ORI: Oceanographic Research Institute; SANBI: South African National Biodiversity Institute; WfW: Working for Water; WSA: Water Services Act.

Source: SWaM, 2019.

4.1.4 Step 4: Design the way forward

Incorporating source-to-sea management into transboundary cooperation requires a shared understanding of the biophysical system and the linkages across the source-to-sea continuum, knowledge of the stakeholders and their roles in relation to source-to-sea challenges, and the strengths and weaknesses of the governance system. Such a shared understanding enables the identification of priorities and the actions needed to achieve goals and objectives. To attain this shared understanding, stakeholders can collectively identify through an inclusive and participatory process a shared vision and the activities required to fulfil this vision, their interrelations and the role of each stakeholder - otherwise known as a theory of change.

Designing a way forward consists of clarifying goals and long-term impacts and then working back from these to identify intervention outputs. It is important to be explicit about causal pathways and assumptions, testing the logic of the latter and assessing the associated risks. Identifying what is within each intervention's scope and defining indicators of success further elaborate the theory of change (GEF/STAP, 2019). Collaborative design of a way forward at the source-to-sea system level can also facilitate a review of individual theories of change prepared by RBOs, RSCAPS, national governments, specific sectors and international donors, among others. This process can support a shift towards greater coherence between actors and levels of governance, thereby strengthening the foundation for system-wide source-to-sea management.

Figure 6 **Theory of change based on four orders of outcomes**

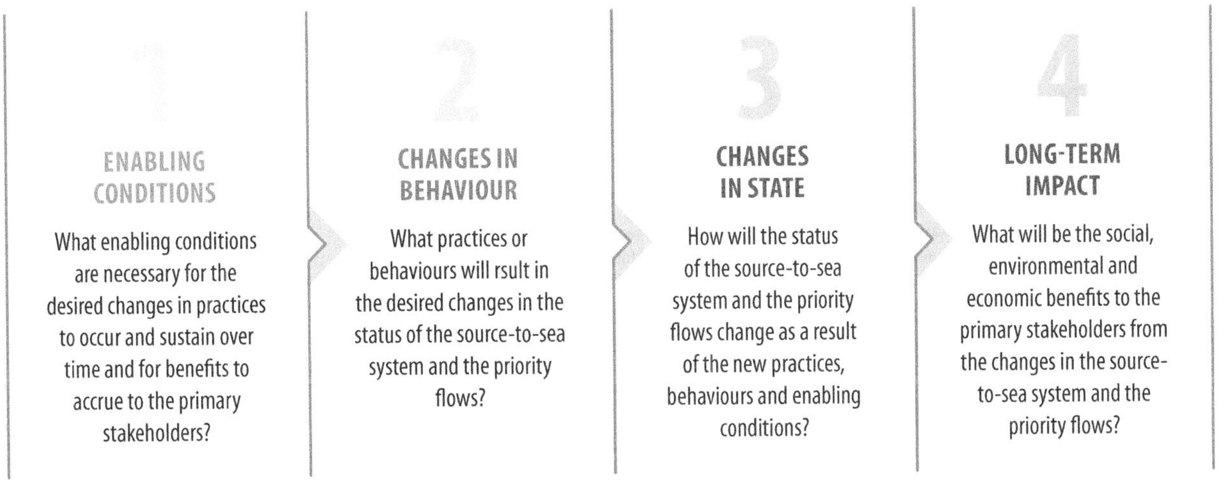

Source: Mathews, Weinberg, Murillo and Liss Lymer, 2023.

Four orders of outcome structure questions to be explored during the design of a theory of change (Figure 6). The first order of outcome is the creation of enabling conditions for source-to-sea management (i.e. institutions and instruments in the governance system facilitate coordination between sectors and strengthen upstream-downstream cooperation). These enabling conditions can range across governance functions, and can include contributions from different institutions and/or instruments. In the transboundary context, institutions responsible for water or regional seas conventions play an important role; however, it is also necessary to address gaps, overlaps and conflicts at the national level.

The second order of outcome is behaviour changes or changes in the practices of resource users and key institutions needed to reduce the sources and impacts of alterations of key flows. The enabling conditions created through the first order of outcome may provide incentives, technical or financial support, or legal requirements for the realization of these changes in behaviour and practices.

The third order of outcome is the achievement of desired changes in societal and environmental conditions across the source-to-sea system. Economic conditions must also be considered for the long-term sustainability of these changes. In the third order of outcome, transboundary cooperation creates benefits for the source-to-sea system as a whole resulting in a resilient source-to-sea system where desired societal and ecological conditions are sustained.

The fourth order of outcome consists of long-term goals and can be related to an idealized future. They can act as a guiding star; however, it is important to set intermediary goals in the first and second order of outcomes that are achievable and measurable.

Figure 7 illustrates a theory of change that uses the four orders of outcomes to identify intervention strategies that will prevent plastic waste leakage into the uMngeni river and its related coastal areas. Their implementation will support the achievement of a sustainable water environment and improved well-being.

Case Study 22 shows how a theory of change can be used to structure project activities. The objective (fourth-order outcome) of the project "Integrating Water, Land and Ecosystems Management in Caribbean Small Island Developing States" was to help preserve Caribbean ecosystems of global significance and improve the sustainability of livelihoods. It worked to reach this goal by reducing environmental stress at project sites in eight countries (third-order outcome). This was achieved by transitioning to sustainable water, land and ecosystems management interventions that account for climate change, which led to enhanced livelihood opportunities and socioeconomic co-benefits for targeted communities (second-order outcome). To this end, the project strengthened policies, monitoring and capacity, and created targeted knowledge-sharing networks (first-order outcome).

Figure 7 **Theory of change showing the first to fourth order of outcomes and intervention strategies for preventing plastic waste leakage**

Intervention Strategies

- Coordination platforms
- Innovative communications platforms
- Trainings
- Workshops
- Awareness building and education
- Public participation meetings
- Market-based instruments for solid waste management and green technnologies

Enabling conditions 1st order outcomes

- Well informed and harmonized leadership in policy and practice
- Higher commitment, accountability and active engagement in solid waste management from source-to-sea
- Improved knowledge and capacities in sustainable watershed management
- Aligned incentives for responsible consumption and production

Changes in behaviour 2nd order outcomes

- Improved coordination and collaboration in policy and practice among stakeholders
- Higher awareness and capacities toward sustainable solutions
- Higher engagement in a progressive power mapping process and decision making processes
- Evaluation of isuues and weaknesses through deliberative processes

Changes in state 3rd order outcomes

Improved governance (transparency, accountability, and participation), reduced environmental stress, and higher social and economic status

Long-term impact 4th order outcomes

Sustainable water environment and improved well-being in uMngeni river lower reaches

Source: SWaM, 2019.

© NN Video / Adobe Stock

22 Case Study

Community engagement in integrating water, land and ecosystems management in Small Island Developing States, Caribbean

Context

The project "Integrating Water, Land and Ecosystems Management in Caribbean Small Island Developing States" (IWEco) was a multi-focal, regional five-year project that built upon the work of previous initiatives, to address water, land and biodiversity resource management as well as climate change in 10 participating countries: Antigua and Barbuda, Barbados, Cuba, the Dominican Republic, Grenada, Jamaica, Saint Kitts and Nevis, Saint Lucia, Saint Vincent and the Grenadines, and Trinidad and Tobago. The project focused on how improved land and water management measures in Small Island Developing States (SIDS) result in reduced stressors, such as pollutants, entering the marine and coastal environment, upon which many Caribbean SIDS are economically dependent.

Project description

The objectives of IWEco were to contribute to the preservation of Caribbean ecosystems of global significance and to help ensure the sustainability of livelihoods through the application of existing proven technologies and approaches appropriate for SIDS through improved fresh and coastal water resources management, sustainable land and forest management, enhancing the resilience of socioecological systems to the impacts of climate change.

The expected outcomes included:

- reduction in environmental stress at project sites in eight countries through appropriate sustainable water, land and ecosystems management interventions that account for climate change;
- enhanced livelihood opportunities and socioeconomic co-benefits for targeted communities;
- strengthened national and regional systems for monitoring of environmental status with respect to key international agreements;
- strengthened national policy and legislation for the effective management of water, land and ecosystems resources that account for climate change;
- strengthened capacity of national and regional institutions and other stakeholders for water, land and ecosystems management that accounts for climate change;
- improved engagement and information access for practitioners and other stakeholders through targeted knowledge-sharing networks.

IWEco was funded by the GEF and implemented by UNEP, the lead implementing agency for national and regional subprojects together with UNDP through the small grant programme. The project was co-executed by the UNEP Cartagena Convention Secretariat and the Caribbean Public Health Agency.

Challenges

Challenges to achieving source-to-sea management included:

- lack of monitoring and sharing of successful approaches among ongoing national subprojects, which would provide new ideas and solutions to common issues as well as improved cross-fertilization;
- lack of coordination between agencies and unclear and inadequate national legal and policy frameworks;
- lack of stakeholder experience on working together to develop strategies and implement integrated solutions across sectors.

Activities

Projects partnered with universities to create opportunities for graduate and postgraduate students to conduct research on the use of nature-based solutions (NBS) and integrated watershed management. In Cuba, for example, universities continued and expanded partnerships begun under the earlier Integrating Watershed and Coastal Areas Management (IWCAM) project. These have resulted in several cohorts graduating from diploma courses, a Master's programme in Integrated Coastal Zone Management (ICZM) and, more recently, a PhD programme. Graduate students conducted trials at farm and reforestation sites on soil health, nutrient recycling, carbon sequestration and biodiversity, and participated in environmental modelling and management planning.

Engaging with school and youth organizations allowed students to visit project sites and practice meaningful field activities. Coverage of these activities led to broader community participation including using project launch events and piggybacking on international and national commemorative event days such as clean ups and tree planting. The well-established "Circles of Interest" in Cuba were the means by which school visits to project sites were facilitated and activities promoted. In Trinidad and Tobago, hundreds of youths from a variety of organizations joined in the tree-planting launch event, which received wide coverage on television and social media.

Lessons learned

The GEF Small Grants Programme (SGP) played a critical role in IWEco by directly supporting local action led by community-based and other civil society organizations in participating countries. The GEF SGP's involvement in IWEco was coordinated by UNDP, which directly disbursed funds in the eight IWEco countries with national subprojects. SGP capacity-building activities leveraged existing stakeholder partnerships and directly engaged with local communities on hands-on activities aligned with respective national subprojects. At their request, activities sought to empower local residents in business development, water management, marine pollution and land restoration.

Source: Case Study provided by Laverne Walker, UNEP Cartagena Convention Secretariat, 2025.

4.1.5 Step 5: Co-create an action plan and foster an enabling environment

Once a theory of change is articulated, attention can focus on identifying intervention strategies to drive progress towards the fourth order of outcomes and co-creation of an action plan. These strategies can complement or coordinate with other relevant plans, whether previous or concurrent, to maximize synergies between them. When determining which stakeholders to involve at each stage of planning and implementation of priority activities, it is helpful to consult the categories of stakeholders in Step 2 (see Table 5). As illustrated in Figure 8, different categories of stakeholders are directly related to specific orders of outcome. Mapping stakeholder relationships to orders of outcome can be helpful when determining the actors to engage in specific intervention strategies and relevant actions.

Success is dependent on both joint and individual actions as well as coordination between actors in the delivery of the action plan. Creating a clear accountability framework of roles and responsibilities, clearly identifying the relationships between actors, will help structure activities and ensure that their timing and synergies are managed beneficially. Indicators for each of the four orders of outcomes should be selected to track progress and can be used to test the underlying assumptions in the theory of change. This can feed into adaptive management as understanding of the linkages across the source-to-sea system improves.

Finance for source-to-sea management can come from a variety of public and private sources, and as understanding of geographic and sectoral interrelationships grows, these sources will diversify. It is important that finance is not limited by economic borders, specific sectors or social geographies, and instead is used to achieve system-wide benefits, taking source-to-sea linkages into consideration, to create the necessary enabling environment for source-to-sea management. Finance mechanisms and enabling environments need to incentivize investment in measures that improve upstream/downstream ecosystems, while assessing the risks and benefits across the source-to-sea continuum. Private philanthropy can also contribute by recognizing the need for source-to-sea management to achieve issue-specific goals and include source-to-sea perspectives in their funding strategies.

Figure 8 **Co-create an action plan and foster an enabling environment**

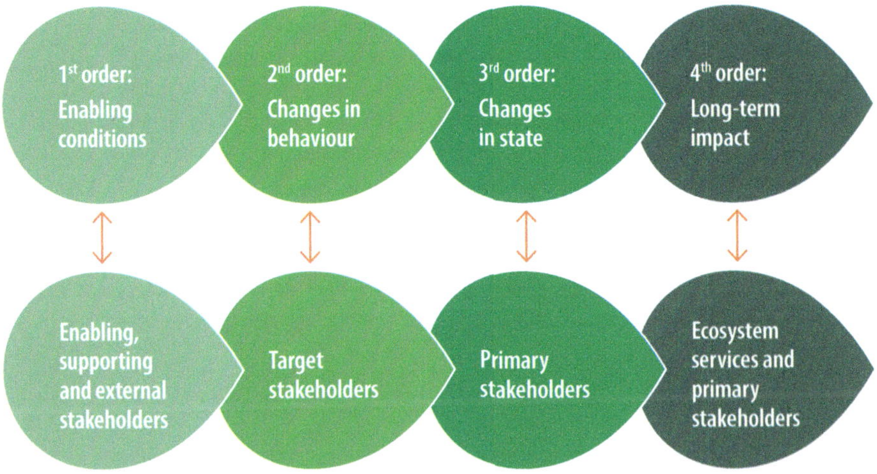

Source: Mathews, Weinberg, Murillo and Liss Lymer, 2023.

International financial institutions such as the GEF are increasingly interested in financing source-to-sea activities and incorporating this emphasis into ongoing transboundary projects. For example, the section on Sustaining Healthy Blue Ecosystems in the GEF-8 Programming Directions supports regional investments that create multi-state cooperation frameworks in transboundary deltas – including an integrated source-to-sea approach. GEF-9 continues support for the source-to-sea approach, which fully aligns with the GEF pillars of integration, whole of society and impact (GEF, 2025).

Funded by the GEF, the Strategic Action Programme for IWRM Management in the Amazon Basin advances source-to-sea management through a range of activities (Case Study 23).

23 Case Study

Introducing source-to-sea cooperation in the Amazon River Basin through a Strategic Action Programme

Context

The Amazon River flows for almost 7,000 km from glaciers high in the Peruvian Andes through the largest tropical forest in the world and eight countries (Bolivia, Brazil, Colombia, Ecuador, Guyana, Peru, Suriname and Venezuela) before discharging its waters in the Atlantic Ocean. Powerful flows of up to 300,000 m^3/s at the mouth result in freshwater advancing almost 200 km into the ocean. Everything in this system is connected from source to sea and through the hydrological cycle.

Pollutants from cities, mining and agricultural runoff flowing into more than 1,000 tributary rivers lead to water quality issues and eutrophication downstream. Deforestation upstream increases soil erosion, impacting water quality and sedimentation patterns, disrupting key ecological processes from rivers to coastal mangroves. Meanwhile, deforestation downstream impacts the Amazon's "flying rivers" reducing rainfall upstream and impacting climate far beyond the basin.

These negative impacts highlight the need for an integrated approach and policy coherence across the source-to-sea system – not only to prevent negative impacts but also to ensure effective achievement of common water and environmental management objectives agreed in the context of the Amazon Cooperation Treaty Organization (ACTO) through a Strategic Action Programme (SAP) for water management and the promotion of ecosystem health, resilience and social development in the region.

Activities

In this context, the Amazon Basin SAP Implementation Project (ACTO/UNEP/GEF) is applying integrated approaches to water management in the Amazon, including integrated water resources management (IWRM), the Water-Energy-Food-Ecosystems (WEFE) Nexus and source-to-sea. The application of source-to-sea management in the Amazon basin supports the establishment of an innovative IWRM governance model based on better understanding of the environmental, social and economic linkages and improved institutional coordination across sectors and segments in the source-to-sea continuum.

Since source-to-sea management is new to the Amazon region, comprehensive training for more than 300 water authorities and professionals in all basin countries was carried out in 2024. The training introduced the approach, illustrating its application in a transboundary river basin context, and identified key entry points to integrate source-to-sea management as part of an IWRM ToolBox specifically designed for the Amazon Region.

Challenges

The implementation of source-to-sea management in the Amazon faces challenges similar to other large river basins, such as fragmented governance and water management along the source-to-sea continuum, insufficient intersectoral and cross-border coordination, absence of regulatory and policy coherence, and a lack of solid and reliable regional data for system-based decision-making. Accordingly, initiatives oriented at overcoming constraints have been promoted in the basin. Constraints specific to the context include insufficient national technical and institutional capacities, conflicting or uncoordinated legislation, insufficient funding for regional projects, and absence of a clear systemic approach to basin management and decision-making. The implementation of source-to-sea management in the Amazon basin is expected to result in more effective resource management, improved adaptation to climate change and ecosystem protection, leading to long-term ecosystem health and resilience and improved quality of life for local populations.

Lessons learned

- Regional Basin Organizations provide space for political and technical dialogue promoting cooperation mechanisms for coordinated water management, both among different sectors and across borders, facilitating the implementation of source-to-sea management.
- Facilitating basin-wide assessments and targeted studies, developing regional strategies and multi-thematic agendas, and promoting shared/common principles and policies, all help advance a systemic vision of the basin from source to sea.
- Institutional capacity building and training are crucial in this process.
- Basin-wide water quality monitoring systems support the reduction of pollution impacts on freshwater and marine system environments along the source-to-sea continuum.

Following the six steps of the source-to-sea approach, the project is working to characterize the source-to-sea system and key flows in the Amazon River Basin. This activity complements the Transboundary Diagnostic Analysis (TDA) previously completed, and will identify and prioritize additional regional strategic interventions, while helping to orient design of the regional monitoring network. In addition, a comprehensive source-to-sea governance assessment will help define the common framework of IWRM principles and guidelines in the Amazon basin, the role of ACTO in water management, and refine the concept for a Permanent IWRM Coordination Mechanism in the Amazon Basin. Finally, a theory of change will provide guidance on governance and management responses, supporting the formulation of additional strategic actions for the SAP and their implementation.

Source: Case Study provided by Maria Apostolova, ACTO, 2025.

© Amazon SAP Implementation Project (ACTO/UNEP/GEF)

Intervention strategies need not be large in scale or complex. In Sweden, the number of horses has increased from 70,000 to approximately 350,000 in 30 years, contributing to nutrients entering water bodies and ultimately the Baltic Sea. To raise awareness and inspire changes in the behaviour of horse owners, the non-profit Race for the Baltic created a targeted communications campaign. By engaging key actors in the sector, they were able to achieve meaningful impact with limited resources (Case Study 24).

24 Case Study

Awareness campaign in Sweden engaging with horse owners to reduce eutrophication in the Baltic Sea

Context

Eutrophication, or the over-enrichment of water from land-based pollutant flows of nitrogen and phosphorus, has severe consequences for ecosystems and the services they provide. For example, in the Baltic Sea this issue can cause toxic algal blooms and oxygen depletion, seriously affecting biodiversity, tourism and fisheries. Sweden is one of the Baltic countries that has been working to reduce flows of these nutrients into water bodies. The country set a target of reducing phosphorus flows by 535 tonnes annually for the period 2007–2021, but fell short of its goal, only managing to reduce phosphorus pollution by 100 tonnes. In Sweden, the number of horses has increased from 70,000 to approximately 350,000 in 30 years, exceeding the number of dairy cows. Today, these horses are kept largely for recreational purposes, often concentrated in peri-urban settings. As a result, horse keeping has the potential to contribute a substantial amount of phosphorus to water bodies in Sweden. Furthermore, horses are regulated differently than other livestock leading to policy gaps, especially on horse manure management. It is estimated that horse manure in Sweden potentially releases around 85 tonnes of phosphorus per year into waterbodies, contributing to the eutrophication of freshwater, coastal and marine ecosystems.

Activities

Up to 70 per cent of phosphorus leakage from horse manure that has been deposited in pastures and paddocks in Sweden can be reduced with simple changes in manure management, such as more frequent mucking, particularly before rainfall or thaws. Based on this knowledge, Race for the Baltic identified an opportunity to support the reduction of phosphorus flows into the Baltic Sea. Together with key stakeholders, the project developed a shared vision, and identified the actions needed to achieve pollution reduction through different activities such as interviews, roundtables and workshops, where knowledge was gathered and shared, including a workshop hosted by Stockholm International Water Institute's Swedish Water House. The key actors consisted of the equestrian division of the Federation of Swedish Farmers (LRF Häst), the Swedish Horse Industry Foundation (HNS), the Swedish Board of Agriculture, and various Water Authorities, County Administrative Boards and independent foundations. The result of this process was a targeted communications campaign.

The aim of the campaign was to raise awareness among horse keepers and riders across Sweden, highlighting the links between horse manure and eutrophication of aquatic ecosystems, and encourage more frequent mucking of pastures and paddocks. To achieve this, the project ran two campaigns in the span of two years using different methods. These included social media campaigns, advertisements in popular media channels dedicated to horse ownership such as magazines, posters sent to more than 8,800 recipients including horseback riding schools, and a dedicated website providing comprehensive information on measures that can be adopted to reduce the negative impact of horse manure on the environment. The communications campaigns were followed up by surveys that measured the adoption of the mucking recommendations. According to the responses received, mucking frequency increased significantly, and horse owners reported being motivated by the understanding that their actions contributed to healthier aquatic ecosystems.

Challenge

Lack of knowledge around the potential contribution of horse manure to nutrient pollution, especially among horse keepers, was the main determinant of activities undertaken by this project. Another significant challenge faced by the project was the need to reach all horse keepers across Sweden, to encourage behaviour change. In order to address this knowledge gap and influence behaviour change, the project assisted with the development of studies on the impacts of horse manure on aquatic ecosystems. This was followed by consultations with key stakeholders from the private and public sector to validate the findings of the studies, and to discuss the best way to approach horse owners. The results of this process were campaigns on different platforms with a country-wide reach.

Concomitant with the launch of the communication campaign, the government released a report on eutrophication that highlighted the potential impact of horse keeping and included mention of several measures that could mitigate its impact – such as restricting the number of horses kept per hectare, further complicating the project. This report created tension among the equestrian community and raised concerns that the project would increase regulations. These concerns were placated through collaboration with stakeholders well known among the horse-keeping community. This resulted in their participation in the communications campaign, among other activities, to demonstrate that nutrient leakage could be achieved through voluntary efforts rather than new regulation.

Results

While the project did not aim to establish regulations on horse manure management, it has created a more favourable environment to implement policies. Through engagement with key stakeholders such as public entities, the private sector and civil society organizations, and the knowledge generated and shared with horse keepers, the project has created a prime space for future policy implementation.

Furthermore, in addition to raising awareness about how lack of horse manure management practices can negatively impact the environment, and how mucking can be an effective measure to mitigate such impacts, the project led to a broader discussion on nutrient capture and cycling within the equestrian community, as well as other ways to reduce and/or capture these resources — including pasture management, horse feed sourcing and quantities, and horse well-being.

This project is an excellent example of implementing far-reaching actions with limited time and resources. The project influenced behavioural change among horse owners, keepers and riders across Sweden, through targeted information campaigns that raised awareness and shared practical recommendations.

Source: Case Study provided by Fanny Tham Ratz, Race for the Baltic, 2025.

Broader intervention strategies can be supported through small projects that pilot innovations. This can be a low investment approach to developing solutions that, if successful, can be upscaled. For instance, to improve water quality in Mtwapa Creek and reduce pollution impacts to downstream coastal and marine ecosystems, a constructed wetland was tested as a form of green infrastructure for wastewater treatment (Case Study 25).

25 Case Study

Improving the water quality of Mtwapa Creek through the use of constructed wetland wastewater treatment technology in Shimo la Tewa, Kenya

Context

Mtwapa Creek, located in Mombasa County, Kenya, experienced significant water quality deterioration due to pollution from various land-based sources. One major contributor was the Shimo la Tewa Prison, which houses between 3,000 and 5,000 individuals and generates approximately 400 m^3 of wastewater daily. The pollution from the prison not only affected Mtwapa Creek but also has downstream impacts on the broader marine ecosystem along Kenya's coastline, influencing water quality and biodiversity beyond national boundaries. Given the interconnected nature of coastal and marine environments, improving wastewater management in the creek contributes to the health of shared regional waters, benefiting both local communities and neighbouring maritime areas.

Activities

This project aimed to improve water quality in Mtwapa Creek by enhancing wastewater management at Shimo la Tewa Prison using a constructed wetland. Constructed wetlands, a form of green infrastructure, are engineered shallow-water ecosystems that mimic natural wetlands to treat wastewater. This project focused on redesigning, rehabilitating and improving the prison's wastewater treatment system into an efficient and low maintenance constructed wetland. By implementing this nature-based solution, the project effectively mitigated point-source pollution from the prison, contributing to the restoration of Mtwapa Creek's and downstream marine water quality.

The project partners were the Kenya Marine and Fisheries Research Institute, Shimo la Tewa Prison, GreenWater – Environmental Water and Sanitation Solutions, and the National Environment Management Authority (NEMA).

The effluent water from the constructed wetland meets NEMA standards, enabling safe water reuse and reducing pollution in Mtwapa Creek. A new water storage system of treated water provides water for landscape and crop irrigation and aquaculture.

However, managing stormwater within the prison complex remained a challenge, as excess water complicated wastewater treatment efforts. Limited technical capacity initially affected the effectiveness of the treatment system, requiring redesigns to ensure optimal function. Despite these hurdles, the project persevered through strong institutional partnerships and adaptive planning.

Results

The innovative technology in the form of a constructed wetland system was effective in the treatment of sewage and wastewater emanating from the prison. Water treated by the constructed wetland is suitable for reuse in greening the environment and in crop production. Beneficiaries of the intervention include the marine ecosystem/environment (improved water quality), the community living near the facility (healthy environment) and improved nutrition (crop production). Effective dissemination of project outcomes through media and research publications has encouraged the broader adoption of constructed wetland technology.

The success of this demo project presents an opportunity to replicate the technology in other correctional facilities and institutions locally and regionally. This is already underway in Mikindani, Kenya, supported by the Go Blue project and coordinated by UNEP. Further dissemination of lessons learned through exchange visits and stakeholder engagement will help scale up source-to-sea management practices, reinforcing long-term environmental benefits.

Source: Case Study provided by Agnes Mukami Muriuki, UNEP, 2025.

© Sam Weru / Caption is Shimo la Tewa terminal review field visit to Shimo la Tewa by NC officials and the terminal evaluation

4.1.6 Step 6: Monitoring, evaluation, learning and adaptation

Monitoring outcomes, capturing and disseminating learning, and adaptively managing for continued success are core elements that can contribute to successful progress towards source-to-sea management. Sharing data and information exchange across sectors and governance boundaries is important for building a comprehensive understanding of the source-to-sea system. This is of particular importance because the impact of actions in one segment of the source-to-sea system on another may not be fully understood unless these relationships are monitored over time. Working toward a shared definition of targets and indicators that reflect stakeholder interests can be an initial step toward source-to-sea cooperation and building partnerships.

Indicators should be selected to monitor the underlying assumptions in the theory of change and the four orders of outcomes elaborated therein (Figure 9). Indicators can be selected to measure:

- successful establishment of enabling conditions (process indicators);
- changes in behaviour and practices used by the targeted stakeholders (stress, or pressure, reduction indicators);
- changes in the status of the source-to-sea system and priority flows (environmental status indicators);
- progress towards the desired long-term impact (impact indicators).

Figure 9 **Indicator types related to the four orders of outcomes in the theory of change**

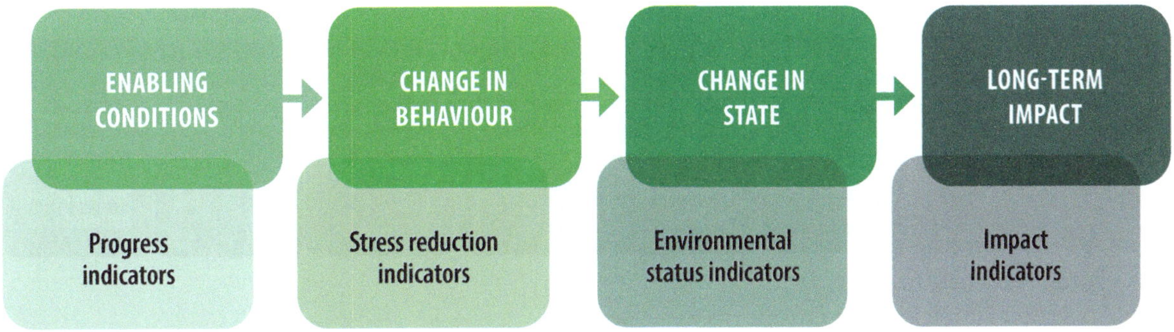

Source: Mathews, Tenberg, Sjödin and Liss Lymer, 2019.

Indicators can also be selected to measure the degree to which source-to-sea management is being adopted by the different stakeholders and whether it has been formalized in governance mechanisms such as laws, policies, procedures and regulations, funding strategies, research agendas and partnership agreements, among others. Monitoring and evaluation of selected indicators should feed into knowledge generation as well as directly into iterative learning cycles through adaptive management. Where unexpected results occur from the implementation of intervention strategies, these can lead to better understanding of the interactions between source-to-sea segments and impact pathways across the source-to-sea continuum. Learning over time leads to a deeper understanding of the source-to-sea system, the stakeholders, the governance system and practices, and may result in adjustments to the intervention strategies.

4.2 Source-to-sea readiness levels

Each regional, national and/or local context has a source-to-sea management readiness level, and the activities undertaken need to reflect the current readiness level (Figure 10).

In the early stages, the focus is raising awareness of source-to-sea linkages and the need for a holistic approach to managing land, freshwater, coasts and the ocean. Existing knowledge of key flows is collected through desk studies and stakeholder consultations, while an initial analysis identifies the stakeholders to engage with in future activities.

During the mid-stages of source-to-sea management readiness, mechanisms for coordination between sectors and across segments are established and the four orders of outcomes are detailed.

At advanced stages, collaborative action planning prioritizes investments in enabling conditions that support changes in behaviour and practices. Stakeholders agree on roles and responsibilities for individual and joint actions. A monitoring system is then put in place to verify the theory of change and deepen the understanding of social, environmental and economic linkages across the source-to-sea continuum. This new knowledge is used to revise the theory of change and to advance further towards source-to-sea management (Table 7).

Progression through the readiness levels can occur over time. The applicability of source-to-sea management at each readiness level is one of its strengths.

Figure 10 Stages of readiness levels for source-to-sea management

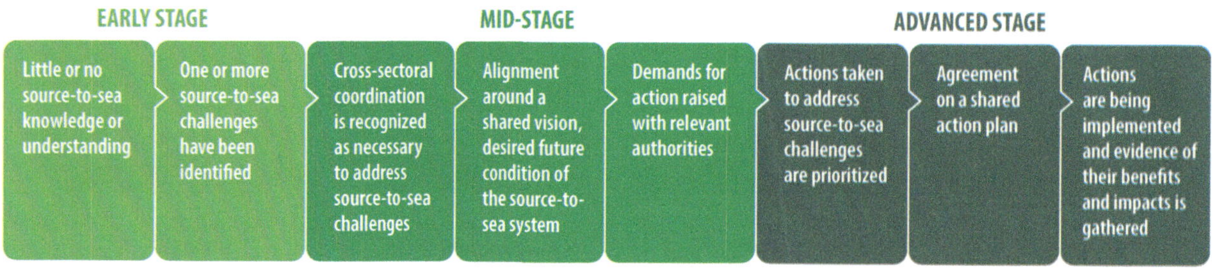

Source: Mathews, Weinberg, Murillo and Liss Lymer, 2023.

© Stephan Röger / Adobe Stock

Table 7 Identifying objectives, activities and outcomes for implementation of the source-to-sea approach based on source-to-sea management readiness levels

Current state of source-to-sea management readiness levels	Key objective	Source-to-sea approach steps to apply	Preparatory activities	Engagement activities	Projected outcome
Little/no source-to-sea knowledge or understanding	Awareness is raised of the linkages from land to freshwater, coasts and the ocean	Step 1: Characterization	Collect existing knowledge of source-to-sea key flows	Share the characterization of key flows from the desk study, consult with stakeholders on knowledge of key flows, and facilitate the selection of priority flows	Deeper understanding of the linkages between land, freshwater, coasts and the ocean from environmental, social and economic perspectives
One or more source-to-sea challenges have been identified	Awareness is raised of the need for a holistic approach to engaging stakeholders across the source-to-system and sectors	Step 2: Engage	Prepare an initial analysis of source-to-sea stakeholders	Share stakeholder analysis with the stakeholders, and facilitate a stakeholder-mapping exercise to feed into the stakeholder engagement plan	Clarification of stakeholders to be engaged in further steps in source-to-sea processes and an engagement plan
Progress on addressing source-to-sea challenges is hampered by lack of cross-sectoral coordination	Capacity is developed and mechanisms established for coordinating strategies, policies, plans and practice across sectors	Step 3: Diagnose	Prepare a governance baseline of institutions and instruments	Share the governance baseline, facilitate the governance-mapping exercise, and conduct training and capacity development in source-to-sea management	Key institutions adopt source-to-sea perspectives and cross-sectoral dialogue is strengthened
Stakeholders need to be aligned around a shared vision of the desired future condition for the source-to-sea system	A shared vision for the future is agreed upon by all stakeholders	Step 4: Design – first stage of identifying the desired future condition	Survey stakeholders on desired future condition	Facilitate the development of a shared vision for the future through an inclusive, bottom-up process	A high-level goal representative of source-to-sea stakeholder interests is broadly agreed upon
Stakeholders' demands for action on source-to-sea challenges have grown and been raised with relevant authorities	Stakeholders agree on a theory of change that expresses how the shared vision for the source-to-sea system will be reached	Step 4: Design – second stage of defining the theory of change	Collate and map steps 1, 2 and 3 outputs to orders of outcomes	Facilitate stakeholders in developing a theory of change that can be used to identify intervention strategies	Assumptions on the linkages between the desired future condition, changes in behaviours and practices, and enabling conditions are explicitly documented

Current state of source-to-sea management readiness levels	Key objective	Source-to-sea approach steps to apply	Preparatory activities	Engagement activities	Projected outcome
Actions to be taken to address the source-to-sea challenge(s) need to be prioritized	Stakeholders agree on the priorities for investments into enabling conditions and support for changing behaviours and practices	Step 5: Act – first stage of prioritizing actions	Prepare a draft assessment of actions	Facilitate priority-setting processes of actions based on intervention strategies developed from the theory of change	A draft plan is prepared with prioritized actions based on an assessment of impact, feasibility and relevance to achieving the desired future
Agreement on an action plan is needed	Agreement is secured from stakeholders on the actions to be taken and roles and the responsibilities to deliver them	Step 5: Act – second stage of accountability for completing actions	Prepare a draft accountability framework for prioritized actions	Facilitate stakeholders in developing an accountability framework for the action plan that clearly indicates roles and responsibilities for each action	A shared understanding of individual and collective responsibilities for implementing the action plan
Actions are being implemented, and evidence of their benefits/impacts is needed	A monitoring system builds an evidence-base of links between alterations in source-to-sea flows, environmental, social and economic impacts, and behaviours, practices and enabling conditions	Step 6: Adapt	Prepare a draft monitoring plan based on theory of change and accountability framework	Engage a monitoring working group to finalize the monitoring plan, establish monitoring and reporting protocols and mechanisms, and secure commitments of long-term investment in monitoring	Long-term monitoring that can feed back into the theory of change and support progressive adaptative management

Source: Mathews, Weinberg, Murillo and Liss Lymer, 2023.

© Raquel Mogado/ Adobe Stock

4.3 Strengths of source-to-sea management

Source-to-sea management offers a holistic and dynamic approach to sustaining ecosystems, balancing human activities and ensuring equitable sharing of benefits. By recognizing the interconnectedness of land, freshwater, coastal and marine environments, this approach strengthens governance and fosters collaboration across sectors and regions. Its effectiveness lies in key principles that shape decision-making, promote inclusivity and enhance resilience. Below are the essential strengths that make source-to-sea management a powerful tool for addressing environmental challenges.

- **Holistic:** This approach recognizes the interconnected nature of ecosystems, ensuring that upstream activities – such as land use, agriculture and industrial processes – are managed in a way that minimizes negative downstream impacts. It integrates environmental, social and economic considerations, fostering sustainable development while balancing costs and benefits across the entire system.
- **Collaborative:** Source-to-sea management builds upon existing institutions and governance frameworks, enhancing coordination among stakeholders. By leveraging established methods and ongoing processes, it strengthens cooperation between sectors such as water management, fisheries and coastal protection, ensuring that efforts are aligned and mutually reinforcing.
- **Prioritizing:** The approach focuses on identifying and addressing the most pressing challenges that have the greatest potential for positive impact. By targeting key issues – such as pollution control, habitat restoration and climate resilience – it maximizes benefits for the entire source-to-sea continuum while minimizing unintended consequences.
- **Inclusive engagement:** Effective source-to-sea management requires active participation from all stakeholders, including marginalized and vulnerable communities. By involving upstream and downstream actors from the outset, it fosters inclusive decision-making and ensures equitable distribution of benefits, strengthening social cohesion and long-term sustainability.
- **Context dependent:** Solutions are tailored to the specific environmental, social and economic conditions of each region. This flexibility allows for adaptive strategies that respond to local challenges, whether addressing urban runoff, agricultural practices or coastal erosion, ensuring that interventions are relevant and effective.
- **Result oriented:** The approach emphasizes measurable outcomes that contribute to broader economic, social and environmental improvements. By setting clear intermediate targets – such as reducing nutrient loads in waterways or restoring critical habitats – it ensures progress toward long-term sustainability goals.
- **Adaptive:** Source-to-sea management embraces a learning-by-doing philosophy, incorporating continuous monitoring and evaluation. This iterative process allows for adjustments based on new data, emerging challenges and evolving stakeholder needs, ensuring resilience and effectiveness over time.

By embracing these strengths, source-to-sea management enables a more integrated, inclusive and effective approach to environmental governance. It ensures that actions taken at every stage of transboundary cooperation contribute to sustainability, resilience and equitable benefit sharing. Through collaboration, prioritization and adaptation, it fosters long-term improvements in ecological health, economic stability and social well-being, creating a lasting impact from source to sea (Case Study 26).

26 Case Study

Strengthening source-to-sea management in the Mediterranean region

Context

Lessons learned from implementation of the MedProgramme indicate that coherent application of source-to-sea management depends on the establishment of a solid governance mechanism to enable enhanced dialogue and coordination among the different actors. Also essential is the development of a sound "theory of change" at the onset of activities (first order), which will drive concerted actions towards the desired impacts (fourth order). The following issues, addressed by the MedProgramme, strengthened source-to-sea management.

Coordination among national actors

Challenge: There is no legal provision or governance mechanism in the partner countries to serve as a framework for coordination on freshwater and marine ecosystem management along the source-to-sea continuum.

Solution: To facilitate inter-institutional coordination and cross-sectoral integration, the MedProgramme built upon existing committees and expanded their mandate from a specific segment of the source-to-sea continuum to multiple segments. In some cases, new coordination committees were created to address a specific integrated approach and then expanded with the aim of bringing together and streamlining all contributions. These coordination mechanisms were established at different levels, either in the field or at a national level to ensure ownership and future sustainability of work.

A careful stakeholder assessment accompanied by meetings and interviews allowed executing agencies to identify the most effective entry point(s) to build a collaborative and constructive collaboration framework among the relevant stakeholders. This coordination process, supported by the MedProgramme Coordinating Unit (MedPCU), forestalled parallel and potentially conflicting unilateral initiatives. Moreover, the MedPCU, through its continued dialogue with national institutions, contributed to the integration of work undertaken by different national actors.

Regional coordination among partner countries

Challenge: National differences in terms of priorities and legal and institutional frameworks pose challenges for dialogue among partner countries.

Solution: Annual Stocktaking Meetings permitted the circulation of information on activities undertaken within the source-to-sea continuum as well as on implementation progress. However, few exchanges among the partner countries were reported. Nevertheless, use of the Medium-Term Strategy and Programme of Work of UNEP/MAP as a reference, together with efforts to develop a regional legal instrument to promote implementation of source-to-sea management, enabled work to proceed towards regional harmonization and possible replication/extension of work in other interested countries.

Lack of an integrated field work component dedicated to source-to-sea pilot implementation

Challenge: Very few provisions were included in the work plan for implementing an actual pilot for source-to-sea solutions.

Solution: The MedProgramme planned activities to improve awareness of source-to-sea management and developed capacities to establish a legal, institutional, technical and financial framework favourable to the implementation of source-to-sea solutions.

Coordination with other donors and agencies

Challenge: Multiple projects were already implemented or under implementation within the water sector in the MedProgramme pilot areas, each with its own approach, methods, scope and involved capacities. Aggregating those multiple initiatives under a constructive multi-player collaboration framework posed several coordination challenges.

Solution: The source-to-sea approach provided a systems reference framework to interconnect, streamline, harmonize and synergize actions across the various projects. The partner countries and their institutions played key roles in clearly identifying priorities and contextualizing experts' recommendations as well integrating efforts made along the different segments of the source-to-sea continuum.

Source: Case Study provided by Mohamad Kayyal and Alessandro Candeloro, UNEP-MAP, 2025.

4.4 Challenges and limitations

While it is generally accepted that the 17 SDGs of the 2030 Agenda for Sustainable Development are indivisible, the communities working toward each SDG and even specific targets under the Goals often operate in silos, with little interaction or coordination with one other. These silos persist in government as well as in civil society, often competing for resources, despite the multitude of environmental, economic and social benefits that different parts of society receive from healthy source-to-sea systems. In some cases, resistance persists to adopting a holistic view.

Source-to-sea systems are complex and there is limited understanding of the social, environmental and economic linkages between segments or how actions in one segment will affect other parts of the system. A complex series of pressures and drivers alter key flows including large-scale processes such as climate change, population growth and economic development. It can be difficult to predict what the outcomes will be across the source-to-sea continuum from actions taken to address these pressures and drivers altering key flows.

Knowledge of source-to-sea linkages and how benefits and costs move both upstream and downstream across the source-to-sea continuum are limited. This can lead to localized actions that are contrary to the aims of holistic management. There may be minimal or no monitoring of changes in the benefits and costs and their sources over time, or there could be limited capabilities for collecting reliable data. Research is often focused on single topics within isolated disciplines with little attention to their intersections within a whole system. Data collection established criteria and the selection of indicators are often not consistent across the different segments of the source-to-sea system, and may be different for each sector. In many cases, it is not feasible to compare and compile data to create a complete picture across segments. Without this possibility, there is no basis for making informed, science-based decisions.

Institutional mandates and jurisdictions are often focused on one sector and/or segment of the source-to-sea system and may not have the capacity to coordinate decisions for benefits at the system level. Examples of good practices, demonstrating the benefits that can come from source-to-sea management, are limited. These challenges can be amplified in cooperation between countries, where there may be inconsistencies among the institutions and instruments responsible for the management of key flows, source-to-sea segments and/or sectors. There may be no single party addressing the source-to-sea system holistically, and no mechanisms to bring stakeholders and public authorities together to support holistic management.

Stakeholders can be geographically distant and come from diverse economic sectors, hold different values or represent conflicting interests. Stakeholder perspectives can often be localized with little recognition of the travel of costs and benefits in both directions of the source-to-sea system. Historically, there have been limited incentives for countries or municipalities to invest in measures that may primarily benefit ecosystems beyond their borders.

Financial systems tend to follow economic borders, often focusing on a specific sector or social geography, thereby limiting the assessment of risks and benefits that could incentivize upstream measures that deliver downstream benefits. Complexities and bureaucracy in accessing existing financial mechanisms can also be a barrier for States and key stakeholders. Without financial mechanisms to bridge actions and benefits, there is limited consideration of the impacts of decisions on other sectors or segments of the source-to-sea system. Investments in implementing source-to-sea management in transboundary cooperation by the Global Environment Facility, as well as others, is providing evidence of its benefits and demonstrating solutions to these challenges.

Chapter 5

Actionable recommendations for accelerating the adoption of source-to-sea management

Key Takeaways:

- Investing in bold, innovative and scalable solutions to source-to-sea challenges multiplies the benefits of cooperation.
- Forging partnerships across all stakeholders supports strategic investments that strengthen cooperation at every level.
- Aligning cross-sectoral policies, mobilizing system-wide investments, fostering innovation and empowering local communities will accelerate transformative actions.
- By establishing effective coordination mechanisms within and across sectors, stakeholders can assess risks and opportunities holistically.
- Catalysing source-to-sea action in transboundary cooperation requires targeted resources, an understanding of implementation bottlenecks, and local, national and regional expertise.
- As insights evolve, potential solutions can be refined and shared across the community, empowering others to navigate challenges with confidence.

Transboundary cooperation is pivotal to tackling the complex challenges that span land, freshwater, coastal and marine ecosystems. When governments, international organizations, NGOs, businesses, academia and citizens unite, they have the power to drive transformative change, creating lasting, positive impacts across source-to-sea systems. Collaboration at global, regional, national and local levels helps prevent biodiversity loss, restore ocean health and protect vital ecosystems from further degradation, while also mitigating the growing threats of climate change. True progress comes from inclusive decision-making, where the voices of marginalized communities, youth, women and Indigenous Peoples are not only heard but actively shape solutions. Investing in bold, innovative and scalable solutions to source-to-sea challenges does not just solve problems, it multiplies the benefits of cooperation, fostering a sustainable future for generations to come.

Embracing source-to-sea management within the framework of transboundary cooperation unlocks powerful, holistic solutions that protect ecosystems, enhance livelihoods and drive global sustainability. Success depends on collaboration across sectors, where governments, NGOs, financial institutions and private enterprises come together as catalysts for change. By forging partnerships, aligning cross-sectoral policies, mobilizing system-wide investments, fostering innovation and empowering local communities, these actors can accelerate transformative actions that create lasting impact – ensuring a resilient future for both people and nature.

Cross-border and cross-sector collaboration is the key to bridging upstream–downstream gaps and ensuring integrated source-to-sea management that delivers lasting benefits across ecosystems and communities. By establishing effective coordination mechanisms within governments and across sectors, stakeholders can assess risks and opportunities holistically, thereby helping to justify strategic investments that strengthen cooperation at all levels.

To truly catalyse source-to-sea management in transboundary cooperation, resources must be directed toward fostering innovative partnerships that unite upstream and downstream stakeholders in solving shared challenges. When diverse actors – from policymakers to local communities – come together, they can transform fragmented efforts into cohesive action that protects ecosystems, strengthens resilience and drives sustainable development across the entire source-to-sea continuum.

Strengthening capacity among public, private and civil society actors is the catalyst for advancing source-to-sea management. By building expertise at local, national and regional levels, stakeholders can effectively tackle upstream-downstream challenges and drive cross-sectoral coordination that fosters sustainable solutions.

To ensure impactful learning and action, a diverse range of tailored capacity-building approaches should be designed, addressing specific source-to-sea challenges while adapting to different contexts and perspectives. Moreover, capacity development must evolve alongside the progress of source-to-sea management, providing the right support at every stage to empower stakeholders and accelerate lasting change.

As source-to-sea management continues to evolve within transboundary cooperation, the key success factors driving its implementation become clearer. Identifying effective strategies for engaging upstream-downstream stakeholders, and fostering cross-sectoral coordination, can strengthen collaborative efforts and accelerate progress.

Understanding the bottlenecks that hinder implementation is equally critical. As insights grow, potential solutions can be explored, refined and shared with the broader community – empowering others to navigate challenges with confidence.

One of the most powerful tools for advancing learning is real-world experience. Collecting and sharing best practices from diverse contexts, and showcasing innovations that deliver tangible results, helps to inspire hesitant stakeholders to take the leap. These examples demystify the process, proving that source-to-sea management is not only achievable but a practical and rewarding approach to tackling water-related challenges and securing long-term environmental and socioeconomic benefits.

Governments, international organizations, NGOs, financial institutions, academic and research institutions, regional and basin organizations, and the private sector, can all contribute to accelerating uptake of source-to-sea management across a broader range of source-to-sea systems. The following lists provide some possible ways forward.

Governments

Governance and policy coordination

- Develop and harmonize policies to address upstream-downstream impacts and link terrestrial, freshwater and marine management to meet both national and regional obligations.
- Integrate source-to-sea principles into water, biodiversity, climate and development policies. Strengthen regulations that align upstream and downstream activities.
- Establish interministerial platforms for cross-sectoral alignment to harmonize actions in agriculture, fisheries, waste management, tourism and coastal planning, among other sectors.
- Facilitate cross-border dialogues to align policies and governance structures across jurisdictions under transboundary cooperation agreements.
- Empower local governments along rivers, coasts and marine areas to collaborate across borders and address shared source-to-sea challenges.
- Develop transboundary agreements where these are lacking.
- Consider sea to source aspects when developing policies, agreements, etc.

Financing and investment

- Create incentives to promote holistic investments that address source-to-sea linkages and generate both local and system-wide benefits.
- Partner with financial institutions to secure funding for transboundary initiatives, especially those fostering cross-sectoral and cross-border collaboration.
- Facilitate PPPs that bring together industries, communities and conservation efforts to implement sustainable practices across the land-freshwater-marine continuum.

Stakeholder engagement

- Empower local governments along rivers and coasts to take active roles in transboundary cooperation, ensuring their concerns are represented in national strategies.
- Facilitate participatory processes that involve Indigenous Peoples, marginalized groups and local communities in planning and decision-making.

Capacity development

- Fund capacity-building programmes for local authorities to enhance understanding and implementation of source-to-sea principles; and support technical training to assess and address transboundary challenges, including pollution control and ecosystem restoration.
- Organize regional forums to share experiences and lessons learned from transboundary source-to-sea initiatives, building bridges between terrestrial, freshwater and marine experts.

International organizations

Governance and policy coordination

- Support countries in developing transboundary agreements that include source-to-sea principles, and establishing joint bodies that facilitate cooperation between RBOs and RSCAPs.
- Advocate for the integration of source-to-sea management into global frameworks like the SDGs, the Paris Agreement and the Convention on Biological Diversity.
- Facilitate transboundary harmonization of national governance frameworks by providing technical support to riparian States.

Financing and investment

- Strengthen existing funding mechanisms, such as the Global Environment Facility, improving accessibility and expediting project development support. Establish new financing channels dedicated to source-to-sea projects.
- Fund studies that explore social, environmental and economic linkages between land, freshwater, coastal and marine ecosystems to strengthen the evidence base for source-to-sea management with a particular focus on transboundary issues.
- Encourage multinational and regional financial cooperation to pool resources for addressing transboundary challenges.
- Promote activities to increase the pool of funds through partnerships, innovations and market designs; and provide assistance to identify and match projects with financial partners.

Knowledge and capacity development

- Offer training programmes to enhance the technical capacities of governments and stakeholders in integrated water and marine resource management.
- Organize knowledge exchange forums to bridge the gap between freshwater and marine actors, fostering cross-sectoral collaboration.

Non-governmental organizations

Stakeholder engagement

- Serve as intermediaries to facilitate dialogue and cooperation between freshwater and marine stakeholders, such as inland water managers and coastal planners.
- Raise awareness through grassroots campaigns about the interconnected impacts of upstream and downstream activities.

Capacity development

- Provide tools and resources to local communities, enabling them to participate effectively in source-to-sea initiatives.
- Work with marginalized groups to empower them with the knowledge needed to influence decision-making processes.
- Facilitate communities of practice for knowledge sharing and create opportunities for peer-to-peer learning. Governments facing similar issues can share their successes, best practices and challenges.

Advocacy and monitoring

- Advocate for system-wide management approaches that emphasize equity, biodiversity conservation and livelihood enhancement.
- Use evidence-based advocacy to push for policies and projects that address the entire source-to-sea continuum, ensuring vulnerable ecosystems are protected.
- Partner with local actors to monitor and evaluate on-the-ground initiatives to assess their effectiveness in connecting freshwater and marine systems.

Financial institutions

Financing and investment

- Develop specific financial instruments and products - such as green bonds or water tariffs - to support sustainable source-to-sea initiatives and encourage businesses and governments to adopt integrated management approaches.
- Establish criteria for funding approvals that require projects to assess and address both upstream and downstream impacts.
- Provide long-term funding for ecosystem restoration projects and prioritize projects that create benefits across the source-to-sea continuum.
- Allocate resources to empower communities and local governments engaged in transboundary cooperation, emphasizing cross-sectoral benefits.
- Create novel market tools, allow for extended grace periods and facilitate guarantees among financial institutions to fund source-to-sea projects.
- Where a project concerns a transboundary water resource with impact on the coastal State, support a commitment to long-term cooperation between the joint body or institution managing the transboundary resource and the coastal State(s).

Capacity development

- Offer financial incentives and capacity-building programmes for private sector actors and communities to adopt sustainable practices.
- Establish platforms or forums to enable financial dialogue among stakeholders.

Private sector

Governance and sustainability

- Align business operations with transboundary environmental standards to minimize downstream, as well as upstream, impacts.
- Collaborate with public and civil society actors to co-develop transboundary initiatives addressing source-to-sea challenges.

Innovation and technology

- Invest in developing and deploying technologies for pollution control, water efficiency and marine debris reduction.
- Support ecosystem services restoration projects, such as wetland conservation and reforestation, to offset operational impacts.

Corporate social responsibility

- Fund community-led restoration projects and capacity-building programmes, contributing to equitable and sustainable outcomes.

Academia and research institutions

Knowledge and evidence building

- Conduct interdisciplinary research to enhance understanding of biophysical, socioeconomic and governance linkages across the source-to-sea continuum.
- Model key flows, such as water balance, sediment transport and pollution dynamics, to inform policy and management decisions.

Capacity development

- Provide education and training programmes for professionals in source-to-sea management.
- Provide education and training programmes on the international legal framework for source-to-sea management, including the Water Convention and others, and on binding international law obligations for States in their utilization of transboundary water resources, or integrate these issues into existing curricula.
- Facilitate science-policy dialogues to translate research findings into actionable recommendations for policymakers.

Regional and Basin Organizations

Governance and policy coordination

- Develop source-to-sea agreements or cooperation between existing river basin and regional seas conventions to address shared transboundary challenges.
- Provide issue-specific guidance for managing water, sediment, biota, pollutants, materials and ecosystem services across borders.
- Operationalize cooperation between RBOs and RSCAPs by developing MOUs, joint working groups or other forms of collaboration.

Stakeholder engagement

- Engage marine and coastal communities in transboundary cooperation in river basins to ensure their concerns are integrated into decision-making processes. Similarly, involve upstream actors in coastal and marine decision-making processes.
- Strengthen partnerships among riparian and coastal States and stakeholders to catalyse collaborative actions.

When governments, international organizations, NGOs, financial institutions, businesses, academia and regional bodies come together, they become the driving force behind effective source-to-sea management in transboundary cooperation. Each player brings something vital to the table – policymaking, financing, research, capacity-building and innovation – all essential for tackling upstream-downstream challenges and shaping holistic solutions. Uniting these efforts allows for the sustainable management of interconnected terrestrial, freshwater, coastal and marine ecosystems, fostering equitable resource sharing, resilient environments, and thriving communities across borders. Such collective action does not just safeguard ecosystems, it paves the way for a future where nature and people flourish in harmony, protecting the source-to-sea continuum for generations to come.

© Iraklis Milas / Adobe Stock

Chapter 6

Conclusions

- **Implementing source-to-sea management in transboundary cooperation will improve outcomes and reap substantial benefits across the source-to-sea continuum.** Source-to-sea management brings added value to transboundary cooperation by ensuring that the social, environmental and economic implications of geographic and sectoral linkages are included and inform decision making. It draws in new stakeholders who bring innovative solutions and incentives for taking source-to-sea action. Transboundary cooperation on source-to-sea management can consolidate resources and reduce costs by focusing on system-wide benefits. It can reduce risk by considering a larger universe of costs and benefits, and can avoid unintended consequences by highlighting impacts that can affect downstream, as well as upstream, ecosystems and communities. The demand for source-to-sea management will increase as awareness of these benefits is raised at local to national and regional levels.

- **Outcomes from transboundary cooperation can be improved by prioritizing source-to-sea challenges and solutions that create optimal benefits for the source-to-sea system as a whole.** Such actions improve the health of freshwater, coastal and marine ecosystems, and generate social, political and economic benefits. Applying a source-to-sea lens to transboundary cooperation and adopting a holistic management of terrestrial, freshwater, coastal and marine systems can prevent biodiversity loss, deterioration of ocean health, ecosystem degradation and exacerbation of climate change impacts.

- **A robust legal foundation for implementing source-to-sea management in transboundary cooperation already exists.** The Water Convention and other global agreements provide a crucial legal and intergovernmental framework for implementing source-to-sea management in transboundary cooperation. Regional agreements, transboundary water agreements, RSCAPs – and the cooperation between them – can facilitate holistic management from source to sea. Existing legal and institutional frameworks for governing shared freshwater resources provide valuable starting points for the sustainable and effective governance of source-to-sea systems, while frameworks for marine water governance add important guidance especially on aims to be reached by addressing impacts from land-based activities.

- **Cooperation on source-to-sea management can take many forms.** Source-to-sea management can advance when States cooperate on the development and implementation of (i) joint action plans, (ii) target setting on pollution prevention, (iii) monitoring programmes, (iv) stakeholder engagement, and (v) planning instruments. These can be facilitated through Joint Task Forces, MOUs, Letters of Agreement and shared Protocols. Institutional arrangements can be implemented according to the context – for example, integration into one instrument (e.g. agreements, policies, protocols, MoU), coordination between different instruments and creating joint commissions.

- **In devising transboundary cooperation, it is important to establish sufficiently strong linkages with national governance frameworks.** To address development challenges from source to sea, effective methods are needed to influence national-level decisions and strategic documents related to the management of source-to-sea segments and key flows. Raising awareness of the benefits of source-to-sea management in meeting global, regional and national obligations on sustainable development, water, the ocean, biodiversity and climate, among others, will strengthen commitment to incorporate source-to-sea perspectives into prominent policy processes. These global priority areas for countries can be tackled through exchanges at higher levels of policy development and actions taken at the local level that contribute to national obligations.

- **Increasing the adoption of source-to-sea management requires commitments from the full range of actors and stakeholders.** Achieving holistic management of land, freshwater, coastal and marine resources that balances environmental and socioeconomic benefits relies on the collective and individual strengths of the community of actors with a stake in source-to-sea challenge(s). Sustainable solutions arise when all stakeholders are included in decision-making, ensuring that the voices and concerns of marginalized and vulnerable communities, youth, women and Indigenous Peoples are heard, and that benefits are shared equitably. Raising awareness of the benefits of source-to-sea management is necessary to secure stakeholder engagement. It is critical that all relevant stakeholders are well-represented as part of the broader knowledge exchange to share in the learning and its applications. It is especially important to include those who have not traditionally been included in decision-making, to ensure they are empowered to participate from an informed position.

- **Resources need to be allocated to foster new forms of collaboration that catalyse source-to-sea management in transboundary cooperation.** Assessing risks and benefits across the source-to-sea continuum can justify finance for coordination between different sectors and tiers of governments, and cooperation across geographic segments. Finance mechanisms and enabling environments need to be created to incentivize investment in measures that address source-to-sea linkages and create benefits for the entire source-to-sea system, while also providing local benefits. Establishing mechanisms that facilitate coordination across levels and sectors of government will reduce conflicts, overlaps and gaps between institutional mandates, and create a coherent governance system. Finance needs to be freed from economic borders, specific sectors or social geographies to enable financing of cross-sectoral and system-wide actions.

- **Demonstrating the potential of source-to-sea management to address development challenges, through context-specific practical applications, will increase action at the ground level.** Building this evidence base will ensure that development on land and along rivers safeguards healthy marine and coastal ecosystems and protects livelihoods dependent upon ecosystem services. Mainstreaming source-to-sea thinking into the design and implementation of projects, and ensuring inclusion in planning processes, governance frameworks and investments will foster new forms of cooperation to address transboundary issues. While financial commitments have been made to source-to-sea projects (e.g. the Global Environment Facility), these need to be substantially expanded.

- **Eliciting commitments from financial institutions to review investments with a source-to-sea lens will accelerate the adoption of source-to-sea management.** Project sponsors, whether international, regional, national or local, should assess project planning, design and implementation through a source-to-sea lens. Opportunities to mainstream source-to-sea thinking in the design and implementation of projects, or promote its inclusion in planning processes, governance frameworks and investments should be pursued.

- **Enhancing knowledge on source-to-sea linkages will strengthen the imperative for and outcomes from source-to-sea management.** The capacity to manage source-to-sea linkages is still limited in most contexts due to gaps in knowledge of these linkages and how impacts and benefits move across the source-to-sea continuum. Investment in science, education, knowledge sharing, data and monitoring is needed to better understand linkages across the source-to-sea continuum. Funding needs to be directed toward research topics that address critical knowledge gaps and can guide policy coherence across the source-to-sea system.

- **Researching areas that create benefits for the whole source-to-sea system can inform science-based policies, management and investments.** Modelling aspects of the source-to-sea system and key flows can deepen understanding of linkages. Data collection, analysis and monitoring needs to be conducted in a manner that provides a coherent view of the source-to-sea system and enables tracking of impacts and benefits.

- **Collectively, innovation in source-to-sea action will lead to transformative, replicable and scalable solutions.** Practical steps can be taken that address source-to-sea challenges while enhancing livelihoods, ensuring equity, harnessing collective wisdom and sustaining ecosystems. Each time a source-to-sea challenge is addressed through transboundary cooperation, accounting for linkages across land, freshwater, coastal and marine ecosystems, new lessons will be learned. Monitoring and evaluation of such source-to-sea action will enable the evidence base on benefits from source-to-sea management to grow.

- **Recognizing the deep interconnectedness of rivers, lakes, aquifers, coastal ecosystems, seas and the ocean is the first step toward holistic management.** These vital water systems transcend national borders, making coordinated efforts among diverse stakeholders essential for protecting ecological health, safeguarding riparian communities, fostering socioeconomic development and ensuring regional stability. Transboundary water cooperation, thus, is the foundation of effective source-to-sea management, allowing nations to work together to address challenges that cross land, freshwater, coastal and marine ecosystems.

- **Existing legal and institutional frameworks for shared freshwater governance offer a strong starting point for sustainably managing entire source-to-sea systems.** Meanwhile, marine governance frameworks provide crucial guidance for mitigating land-based impacts on coastal and ocean ecosystems. Embracing source-to-sea management in transboundary cooperation, therefore, will enable the world's precious water resources to be managed sustainably for generations to come.

References and supporting resources

Berggren, Jens, and Birgitta Liss Lymer (2016). Source to Sea: Linkages in the 2030 Agenda for Sustainable Development. Stockholm: Swedish Agency for Marine and Water Management.

Birnie, Patricia W., Alan Boyle, and Catherine Redgwell (2009). *International Law and the Environment.* Third edition. Oxford University Press. Available at: http://oceanlaw.ru/wp-content/uploads/2017/10/International-Law-and-the-Environment-Third-Edition-Patricia-Birnie.pdf

Crossland, Christopher. J., Hartwig H. Kremer, Han J. Lindeboom, Janet I. Marshall Crossland, and Martin D.A. Tissier, eds (2005). *Coastal Fluxes in the Anthropocene: The Land-Ocean Interactions in the Coastal Zone Project of the International Geosphere-Biosphere Programme.* Berlin, Springer Berlin-Heidelberg, https://doi.org/10.1007/3-540-27851-6

Elliot, Michael, Ángel Borja, and Roland Cormier (2023). Managing marine resources sustainably – Ecological, societal and governance connectivity, coherence and equivalence in complex marine transboundary regions. *Ocean and Coastal Management*, vol. 245, 106875.

Elliott, Michael, and Alan K. Whitfield (2025). Science and management achieving connectivity, coherence and equivalence to ensure the health of estuarine fish communities. *Estuarine, Coastal and Shelf Science*, vol. 314, 109133.

Global Environment Facility (2019). *Theory of Change Primer.* GEF/STAP/C.57/Inf.04. Washington, D.C. Available at https://www.thegef.org/sites/default/files/council-meeting-documents/EN_GEF_STAP_C.57_Inf.04_Theory%20of%20Change%20Primer_0.pdf

__________ (2023). *Incorporating Co-Benefits in the Design of GEF Projects.* GEF/STAP/C.64/Inf.03. Washington, D.C. Available at https://www.thegef.org/sites/default/files/documents/2023-06/EN_GEF.C.64.STAP_.Inf_.03_Incorporating_cobenefits_in_the_design_of_GEF_projects.pdf

__________ (2024). *Fostering Cooperation and Managing Conflict: A STAP Information Note on GEF Transboundary Water Projects.* GEF/STAP/C.67/Inf.04. Washington, D.C. Available at https://www.thegef.org/council-meeting-documents/gef-stap-c-67-inf-04

Global Environment Facility, and World Bank Group (2025). *Ninth GEF Replenishment: Draft GEF-9 Strategic Positioning and Programming Directions.* GEF/R.9/05. Washington, D.C.: GEF Secretariat. Available at https://www.thegef.org/sites/default/files/documents/2025-04/GEF.R.9.05-%20Draft%20GEF-9%20Strategic%20Positioning%20and%20Programming%20Directions_0.pdf

Granit, Jakob, Birgitta Liss Lymer, Stephen Olsen, J. Lundqvist, and A. Lindstrom (2014). *Water Governance and Management Challenges in the Continuum from Land to the Coastal Sea – Spatial Planning as a Management Tool.* SIWI Working Paper No. 22. Stockholm: Stockholm International Water Institute.

Granit, Jakob, Birgitta Liss Lymer, Stephen Olsen, Anne Tengberg, Sulev Nõmmann, and Torkil Jønch Clausen (2017a). *A Conceptual Framework for Governing and Managing Key Flows in a Source-to-Sea Continuum: A STAP Advisory Document.* Washington, D.C.: Global Environment Facility. Available at https://www.thegef.org/sites/default/files/publications/S2S%20conceptual%20framework_web%20version.pdf

Granit, Jakob, Birgitta Liss Lymer, Stephen Olsen, Anne Tengberg, Sulev Nõmmann, and Torkil Jønch Clausen (2017b). A conceptual framework for governing and managing key flows in a source-to-sea continuum. *Water Policy*, vol 19, no. 4, pp. 673–991. https://doi.org/10.2166/wp.2017.126

Groeneweg-Thakar, Kanika, Ruth E. Mathews, Birgitta Liss Lymer, and Victor Sundman (2020). *Starting at the Source to Save The Sea: Look Upstream to Achieve SDG 14.* Stockholm International Water Institute for the Action Platform for Source-to-Sea Management, Stockholm, Sweden. Available at https://siwi.org/wp-content/uploads/2020/06/starting-at-the-source-to-save-the-sea-final_webb.pdf

Hebart-Coleman, David, and Ruth E. Mathews (2023). *Building Urban Resilience through Strategic and Environmental Planning from Source to Sea.* Policy Brief. Stockholm: Stockholm International Water Institute. Available at https://siwi.org/wp-content/uploads/2023/01/building-urban-resilience-from-sourceto_sea-policy-brief-final.pdf

International Union for Conservation of Nature (2010). *Integration of Biodiversity in the Decision-Making Process: Lessons Learnt from the Moulouya Projects.* Gland, Switzerland. Available at https://iucn.org/sites/default/files/import/downloads/moulouya_lessons_learned.pdf

Integrating Water, Land and Ecosystems Management in Caribbean Small Island Developing States (2024). *Cross-cutting Case Study for Common Barriers and Solutions to Integrated Land, Water and Ecosystems Management in Caribbean SIDS of the IWEco Project.* Available at https://iweco.org/sites/default/files/2024-06/IWEco_Cross-Cutting_Case_Study_Final_2024.pdf

Jimenez, Alejandro, Panchali Saikia, Richard Giné, Pilar Avello, James Leten, Birgitta Liss Lymer, Kerry Schneider, and others (2020). Unpacking water governance: A framework for practitioners. *Water*, vol. 12, no. 3, 827. https://doi.org/10.3390/w12030827

Kellock, Celeste, Chris Leakey, Christian Schröder, Nicola Melville, Janet Khan, and Scot Mathieson (2023). *Source to Sea – Enabling Coherent, Efficient and Synergistic Outcomes*. NatureScot Research Report 1343. Inverness, Scotland, Nature Scot. Available at https://www.nature.scot/doc/naturescot-research-report-1343-source-sea-enabling-coherent-efficient-and-synergistic-outcomes

Mathews, Ruth. E., and Joachim Stretz (2019). *Source-to-Sea Framework for Marine Litter Prevention: Preventing Plastic Leakage in River Basins*. Stockholm: Stockholm International Water Institute. Available at https://siwi.org/wp-content/uploads/2019/08/marine-litter-prevention_2019_webb-1.pdf

Mathews, Ruth. E., Anna Tenberg, Johanna Sjödin, and Birgitta Liss Lymer (2019). *Implementing the Source-to-Sea Approach: A Guide for Practitioners*. Stockholm: Stockholm International Water Institute. Available at https://siwi.org/wp-content/uploads/2024/01/source-to-sea-guide_webb.pdf

Mathews, Ruth. E., Josh Weinberg, Jose Murillo, and Birgitta Liss Lymer (2023). *Building Momentum to Accelerate Adoption of Source-To-Sea Management: Lessons Learned and Recommendations from Seven Case Studies*. Gothenburg, Sweden: Swedish Agency for Marine and Water Management. Available at https://siwi.org/wp-content/uploads/2023/09/rapport-2023-8-building-momentum-to-accelerate-adoption-of-source-to-sea-management.pdf

Michels-Brito, Adriane, José Carlos Ferreira, and Carlos Hiroo Saito (2023). Source-to-sea, integrated water resources management, and integrated coastal management approaches: Integrative, complementary, or competing? *Journal of Coast Conservation*, vol. 27, no. 66. https://doi.org/10.1007/s11852-023-00999-z

__________ (2024). The source-to-sea landscape: A hybrid integrative territory management approach. *Science of the Total Environment*, vol. 931, 17296. https://doi.org/10.1016/j.scitotenv.2024.172961

Olsen, Stephen. B. (2003). Frameworks and indicators for assessing progress in integrated coastal management initiatives. *Ocean & Coastal Management*, vol. 46, pp. 347–361. https://doi.org/10.1016/S0964-5691(03)00012-7

Open Plan (2021). *Review and Strengthen Policy, Legislative and Institutional Capacity to Support Sustainable Land and Water Resources and Ecosystems Management in IWEco Participating States*. Integrated Management System Assessment Model. Lincoln. UK. Available at https://iweco.org/sites/default/files/2022-06/IWEco_Project_C3_Integrated_Management_System_Assessment_Model_Final_June2021.pdf

Programme in Water Conflict and Transformation (2025a). International Freshwater Treaties Database. Corvallis, OR: Oregon State University. Available at https://transboundarywaters.ceoas.oregonstate.edu/international-freshwater-treaties-database. Accessed on 1 September 2025.

Programme in Water Conflict and Transformation (2025b). International River Basin Organization Database. Corvallis, OR: Oregon State University. Available at https://transboundarywaters.ceoas.oregonstate.edu/international-river-basin-organization-database. Accessed on 1 September 2025.

Ramesh, R., Z. Chen, V. Cummins, J. Day, C. D'Elia, B. Dennison, D.L. Forbes, and others (2015). Land–ocean interactions in the coastal zone: Past, present & future. *Anthropocene*, vol. 12, pp. 85–98. https://doi.org/10.1016/j.ancene.2016.01.005

S2S Platform (2021). *Action Platform for Source-to-Sea Management: Strategy 2021–2025*. Stockholm: Stockholm International Water Institute. Available at https://siwi.org/wp-content/uploads/2021/01/action-platform-for-source-to-sea-management-strategy-2021-2025.pdf

Swedish Agency for Marine and Water Management (2019). *Source-to-Sea – Screening study of Pollution and Flows in KwaZulu-Natal, South Africa*. NIRAS International Consulting with GroundTruth, South Africa.

Swedish International Development Cooperation Agency (2022). *Applying a Source to Sea Approach in Sida's Development Cooperation*. Available at https://cdn.sida.se/app/uploads/2022/03/08104227/10205932_Sida_Brief_Source_to_sea_feb-2022_webb.pdf

United Nations Economic Commission for Europe (2013a). *Guide to Implementing the Water Convention*. United Nations publication. Available at https://unece.org/DAM/env/water/publications/WAT_Guide_to_implementing_Convention/ECE_MP.WAT_39_Guide_to_implementing_water_convention_small_size_ENG.pdf

__________ (2013b). *Guide to Public Participation under the Protocol on Water and Health*. United Nations publication. Available at https://unece.org/DAM/env/water/publications/PWH_public_participation/GuidePublicParticipationPWH_WEB_EN.pdf

___________ (2015). Policy Guidance Note on the Benefits of Transboundary Water Cooperation: Identification, Assessment and Communication. United Nations publication. Available at https://unece.org/DAM/env/water/publications/WAT_47_Benefits/ECE_MP.WAT_47_PolicyGuidanceNote_BenefitsCooperation_1522750_E_pdf_web.pdf

___________ (2021). *Practical Guide for the Development of Agreements or Other Arrangements for Transboundary Water Cooperation.* United Nations publication. Available at https://unece.org/sites/default/files/2021-11/ece_mp.wat_68_eng.pdf

___________ (2024a). *Progress Report on Transboundary Water Cooperation under the Water Convention, Third Report on the Implementation of the Convention on the Protection and Use of Transboundary Watercourses and International Lakes 2020–2023.* United Nations publication. Available at https://unece.org/sites/default/files/2024-12/2417627_E_PDF_WEB.pdf

___________ (2024b). *Convention on the Protection and Use of Transboundary Watercourses and International Lakes: Draft Programme of Work 2025–2027.* Available at https://unece.org/sites/default/files/2024-05/Item%2015_PoW2025-2027.pdf

United Nations Economic Commission for Europe, United Nations Educational, Scientific and Cultural Organization, and UN-Water (2024). *Progress on Transboundary Water Cooperation: Mid-term status of SDG Indicator 6.5.2, with a special focus on Climate Change – 2024.* United Nations publication. Available at https://www.unwater.org/sites/default/files/2024-09/SDG6_Indicator_Report_652_Progress-on-Transboundary-Water-Cooperation_2024_EN.pdf

UN-Water (2020). *UN-Water Policy Brief on the United Nations Global Water Conventions: Fostering Sustainable Development and Peace.* United Nations publication. Available at https://www.unwater.org/sites/default/files/app/uploads/2021/01/UN-Water_Policy_Brief_United_Nations_Global_Water_Conventions.pdf

___________ (2022). *High-Level Symposium on Water: Bridging SDG 6 and SDG 14.* Special Event of the United Nations Conference. Available at www.unwater.org/news/high-level-symposium-water-2022-un-ocean-conference

Printed and bound by CPI Group (UK) Ltd, Croydon, CR0 4YY
09/11/2026
14999543-0005